Genetic Engineering

Principles and Methods

Volume 28

GENETIC ENGINEERING
Principles and Methods

A Continuation Order Plan is available for this series. A continuation order will bring delivery of each new volume immediately upon publication. Volumes are billed only upon actual shipment. For further information please contact the publisher.

Genetic Engineering

Principles and Methods

Volume 28

Edited by

Jane K. Setlow
Brookhaven National Laboratory
Upton, New York

Springer

The Library of Congress cataloged the first volume of this title as follows:

Genetic engineering: principles and methods. V. 1–
New York, Plenum Press. (1979–
v. ill. 26 cm.
Editors: J. K. Setlow and A. Hollaender
Key title: Genetic engineering. ISSN 0196-3716

1. Genetic engineering—Collected works. I. Setlow, Jane K. (1979–) II. Hollaender, Alexander, (1979–1986).

QH442.G454 575.1 76-644807
MARC-S

ISBN-10: 0-387-33840-3 e-ISBN-10: 0-387-34504-3
ISBN-13: 978-0-387-33840-8 e-ISBN-13: 978-0-387-34504-8

Printed on acid-free paper.

9 8 7 6 5 4 3 2 1

springer.com

ACKNOWLEDGMENT

The Editor again praises Bonnie McGahern for the competent final processing, including fixing some of the Editor's errors.

CONTENTS OF EARLIER VOLUMES

VOLUME 7 (1985)

VOLUME 8 (1986)

VOLUME 14 (1992)

VOLUME 15 (1993)

VOLUME 16 (1994)

VOLUME 17 (1995)

VOLUME 21 (1999)

Nuclear Plasmids of *Dictyostelium* • *Joanne E. Hughes and Dennis L. Welker*

The Translation Initiation Signal in *E. coli* and Its Control • *Eckart Fuchs*

Direct Isolation of Specific Chromosomal Regions and Entire Genes by Tar Cloning • *Vladimir Larionov*

Regulation of Lysine and Threonine Metabolism in Plants • *Rachel Amir and Gad Galili*

Genetic Engineering of Plant Chilling Tolerance • *James Tokuhisa and John Browse*

Role of Bacterial Chaperones in DNA Replication • *Igor Konieczny and Maciej Zylicz*

Transformation of Cereals • *Roland Bilang, Johannes Fütterer, and Christof Sautter*

Mechanisms of Initiation of Linear DNA Replication in Prokaryotes • *Margarita Salas*

Diverse Regulatory Mechanisms of Amino Acid Biosynthesis in Plants • *Katherine J. Denby and Robert L. Last*

Forage and Turf-Grass Biotechnology: Principles, Methods, and Prospects • *John W. Forster and Germán C. Spangenberg*

Informatics Needs of Plant Molecular Biology • *Mary Polacco*

VOLUME 22 (2000)

Post-Transcriptional Light Regulation of Nuclear-Encoded Genes • *Marie E. Petracek and William F. Thompson*

Novel Methods of Introducing Pest and Disease Resistance to Crop Plants • *Jeremy Bruenn*

Targeting Gene Repair in Mammalian Cells Using Chimeric Oligonucleotides • *Eric B. Kmiec, Sarah Ye, and Lan Peng*

Exploring the Mechanism of Action of Insecticidal Proteins by Genetic Engineering Methods • *Jeremy L. Jenkins and Donald H. Dean*

Enzyme Stabilization by Directed Evolution • *Anne Gershenson and Frances H. Arnold*

ET-Cloning: Think Recombination First • *Joep P. P. Muyrers, Youming Zhang, and A. Francis Stewart*

Growth and Genetic Modification of Human β-Cells and β-Cell Precursors • *Gillian M. Beattie, Albert Hayek, and Fred Levine*

Elucidation of Biosynthetic Pathways by Retrodictive/Predictive Comparison of Isotopomer Patterns Determined by NMR Spectroscopy • *Wolfgang Eisenreich and Adelbert Bacher*

Are Gene Silencing Mutants Good Tools for Reliable Transgene Expression or Reliable Silencing of Endogenous Genes in Plants? • *Philippe Mourrain, Christophe Béclin, and Hervé Vaucheret*

Manipulating Plant Viral RNA Transcription Signals • *Cynthia L. Hemenway and Steven A. Lommel*

Genetic Engineering Strategies for Hematologic Malignancies • *Thomas J. Kipps*

Telomerase and Cancer • *Murray O. Robinson*

VOLUME 23 (2001)

Evolution of Transport Proteins • *Milton H. Saier, Jr.*

Mechanisms of Apoptosis Repression • *Collin C. Q. Vu and John A. Cidlowski*

Cytokine Activation of Transcription • *Kerri A. Mowen and Michael David*

Enzymatic Approaches to Glycoprotein Synthesis • *Pamela Sears, Thomas Tolbert, and Chi-Huey Wong*

Vector Design and Development of Host System for *Pseudomonas* • *Herbert P. Schweizer, Tung T. Hoang, Katie L. Propst, Henry R. Ornelas, and RoxAnn R. Karkhoff-Schweizer*

Genetic and Biochemical Studies on the Assembly of an Enveloped Virus • *Timothy L. Tellinghuisen, Rishika Perera, and Richard J. Kuhn*

Characterization of Protein Structure and Function at Genome Scale Using a Computational Prediction Pipeline • *Dong Yu, Dongsup Kim, Phuongan Dam, Manesh Shah, Edward C. Uberbacher, and Ying Xu*

VOLUME 26 (2004)

Arabidopsis as a Genetic Model for Interorganeille Lipid Trafficking • *Christoph Benning, Changcheng Xu, and Koichiro Awai*

Protein Sequence Database Methods • *Maria Jesus Martin, Claire O'Donovan, and Rolf Apweiler*

Properties and Applications of Cell-Penetrating Peptides • *A. Gräslund and L.E.G. Eriksson*

Detection of Topological Patterns in Protein Networks • *Sergei Maslov and Kim Sneppen*

Dna Microarrays: Methodology, Data Evaluation and Applications in the Analysis of Plant Defense Signaling • *E. Kuhn and A. Schaller*

Approaches for Identification of Fungal Genes Essential for Plant Disease • *Candace E. Elliott and Barbara J. Howlett*

Genetic Mapping in Forest Trees: Markers, Linkage Analysis and Genomics • *Matias Kirst, Alexander Myburg, and Ronald Sederoff*

The Production of Long Chain Polyunsaturated Fatty Acids in Transgenic Plants • *Johnathan A. Napier, Frédéric Beaudoin, Louis V. Michaelson, and Olga Sayanova*

Investigating *in situ* Natural Genetic Transformation of *Acinetobacter* Sp. BD413 in Biofilms with Confocal Laser Scanning Microscopy • *Larissa Hendrickx and Stefan Wuertz*

The Path in Fungal Plant Pathogenicity: Many Opportunities to Outwit the Intruders? • *Guus Bakkeren and Scott Gold*

Analysis and Annotations of Microbial Genome Sequences • *Loren Hauser, Frank Larimer, Miriam Land, Manesh Shah, and Ed Uberbacher*

Brain Plasticity and Remodeling of Ampa Receptor Properties by Calcium-Dependent Enzymes • *Guy Massicotte and Michel Baudry*

Gene Regulation by Tetracyclines • *Christian Berens and Wolfgang Hillen*

VOLUME 27(2005)

Identification and Analysis Of Micrornas • *Sveta Bagga and Amy E. Pasquinelli*

Dormancy and the Cell Cycle • *Michael A. Campbell*

The Opt Family Functions in Long-Distance Peptide and Metal Transport in Plants • *Mark Lubkowitz*

Phospholipid-derived Signaling in Plant Response to Temperature and Water Stresses • *Xuemin Wang*

Anionic Nutrient Transport in Plants: The Molecular Basis of the Sulfate Transporter Gene Family • *Hideki Takahashi, Naoko Yoshimoto, and Kazuki Saito*

Purification of Protein Complexes by Immunoaffinity Chromatography: Application to Transcription Machinery • *Nancy E. Thompson, Debra Bridges Jensen, Jennifer A. Lamberski, and Richard R. Burgess*

Biogenesis of Iron-Sulfur Cluster Proteins in Plastids • *Marinus Pilon, Salah E. Abdel-Ghany, Douglas Van Hoewyk, Hong Ye, and Elizabeth A. H. Pilon-Smits*

Iron Transport and Metabolism in Plants • *Loubna Kerkeb and Erin L. Connolly*

Salt Stress Signaling and Mechanisms of Plant Salt Tolerance • *Viswanathan Chinnusamy, Jianhua Zhu, and Jian-Kang Zhu*

Strategies for High-throughput Gene Cloning and Expression • *L. J. Dieckman, W. C. Hanly, and F. R. Collart*

Molecular Roles of Chaperones in Assisted Folding and Assembly of Proteins • *Mark T. Fisher*

Engineering Plants for Increased Nutrition and Antioxidant Content Through the Manipulation of the Vitamin E Pathway • *David K. Shintani*

CONTENTS

REGULATION OF PLANT INTERCELLULAR COMMUNICATION VIA PLASMODESMATA

Insoon Kim, Ken Kobayashi, Euna Cho and Patricia C. Zambryski

Department of Plant and Microbial Biology
111 Koshland Hall
University of California
Berkeley, CA 94720

INTRODUCTION

Plasmodesmata (PD) are unique to plants, and are utilized to establish dynamic intercellular continuity between groups of cells enabling the transport of nutrients, developmental cues and ribonucleoprotein complexes (reviewed in (1-4)). Multidisciplinary investigations over the last decade provide evidence that plasmodesmatal regulation is critical to various basic plant functions such as development, host–pathogen interactions and systemic RNA silencing. This chapter highlights various tools used to study PD, and elaborates on the regulation of PD during plant development.

PLASMODESMATA: STRUCTURES AND FUNCTIONS

Generic simple PD have two major components, membranes and spaces (5) (Figure 1). Membranes constitute boundaries of the PD channel. The plasma membrane (PM) of two neighboring cells form the outer boundary of PD.

Genetic Engineering, Volume 28, Edited by J. K. Setlow

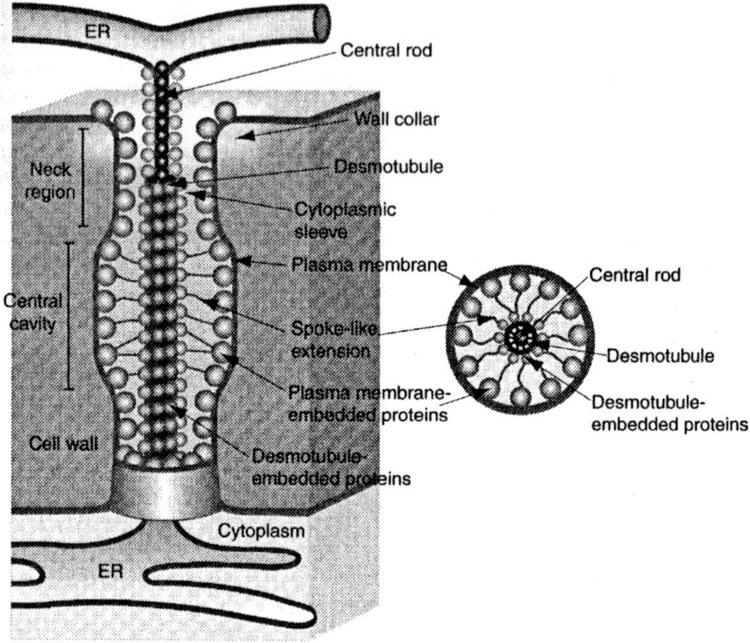

Figure 1. Diagram of simple PD. A longitudinal view in the left and a transverse view in the right, reprinted with permission from (5).

Appressed endoplasmic reticulum (ER), termed the desmotubule (D), runs through the axial core of PD and forms the inner boundary. The space between PM and D is the cytoplasmic sleeve (CS), the primary passageway for molecular transport, which is continuous with the cytoplasm between adjacent cells. The CS is not empty. Instead, the CS is filled with proteinaceous molecules that likely regulate transport *via* PD. For example, actin and myosin along the length of PD (reviewed in (6)), and centrin nanofilaments at the neck region (7), may provide contractile elements to control PD apertures.

The functional measure of PD is their size exclusion limit (SEL), the upper limit of the size of macromolecules that can freely diffuse from cell to cell. PD SEL is regulated temporally, spatially and physiologically throughout plant development. PD selectively allows movement of proteins, such as transcription factors, and RNAs, such as mRNAs and silencing RNAs, both critical in cell-fate determination (reviewed in (8, 9)). Therefore, PD in different tissues may be regulated differentially, possibly by the involvement of developmentally-regulated factors.

When cells and tissues exhibit cell-to-cell transport of micro- or macromolecular tracers they are said to form “symplastic domains” of shared cytoplasm. Cells within symplastic domains share a common PD aperture (SEL) compared to cells in surrounding regions. Because symplastic domains are thought to form during differentiation of tissues/organs, the determination of which cells and tissues in the plant are in communication *via* PD is an area of active investigation.

Such studies reveal communication domains for developmental/morphogenetic signaling. Below we review PD function during adult, seedling and embryonic plant development.

PD Function During Adult Plant Development

Research on PD has made exponential progress in the last several years due to technical innovations. The major targets of PD research in adult plants are leaves (see Figure 2 for plant diagram), due to their ready accessibility. The first approach used to examine PD function was microinjection of fluorescent probes. Historically, this approach revealed that PD SEL was less than 1 kDa (10), and only few specialized viral (11) or homeodomain proteins (12) could dilate PD beyond their innate small apertures.

The use of green fluorescent protein (GFP) and its introduction by biolistic bombardment dramatically altered this view, revealing an inherent complexity

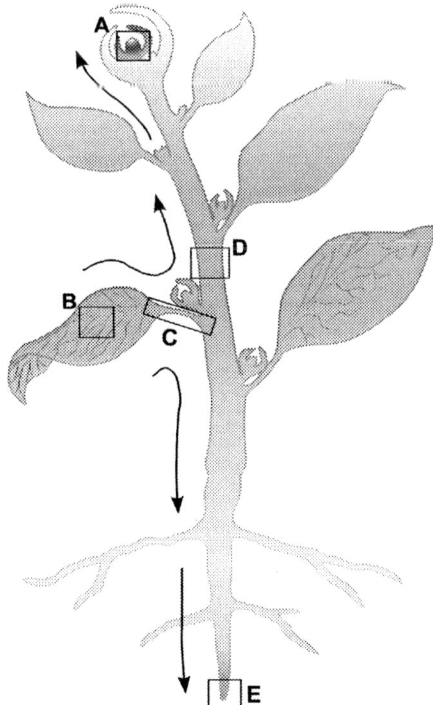

Figure 2. Plants use a combination of local and long-distance signaling to orchestrate proper function throughout the whole plant. Signals perceived/generated by leaves (B) are transmitted along the vascular systems of petioles (C) and stems (D), and then delivered to distant organs such as the shoot apical meristem (A) and the root tip (E). Signals generated in leaves and transported through the phloem reflect environmental changes (light, temperature, mineral nutrients, and water availability), physiological programs, developmental cues, and pathogenic attacks. Arrows indicate the transport by the phloem of the vascular system. Adapted from (3) with permission.

of PD function. Basically, plant leaves are bombarded with DNA constructs to express GFP (27 kDa) or its larger-sized protein fusions. Remarkably, such studies revealed that proteins at least 50 kDa were able to traffic cell-to-cell by passive diffusion (13, 14). PD aperture in leaves is developmentally regulated. Younger leaf cells contain PD with more dilated aperture than older leaves (as measured by different-sized GFP tracers) and this function is correlated with structural changes in PD that occur during leaf maturation. Quantitative studies, one of the benefits offered by biolistic bombardment over microinjection, reveal that even a single leaf is composed of PD with various apertures that likely respond dynamically to environmental and physiological changes (15).

PD aperture and protein size obviously govern passive macromolecular traffic. Given that size and aperture are synchronous with each other, can all macromolecules move cell-to-cell like GFP? Such rampant exchange would lead to loss of critical cell components. By fusing GFP to several localization sequences, such as ER retention or cytoskeleton anchoring, it was determined that cellular location dictates whether or not a protein can move cell-to-cell (14). Thus, exogenous tracers such as GFP can move by default as they do not contain cellular targeting signals. However, cells likely sequester or anchor their proteins according to their functions and thereby protect against non-specific intercellular transport.

Another method to measure PD conductivity, phloem loading, takes advantage of the plant vascular system. Fluorescent membrane impermeable tracers (once they are in the cytoplasm, they can move cell-to-cell only *via* PD) are loaded from the end of cut petioles, the little branch remaining after removal of leaves (see Figure 2 for plant parts). Tracers load into and move along the phloem. Tracers can then "unload" *via* PD connections between the phloem and surrounding cells in sink leaves or at the shoot apex. As tracer movement is imaged at a distance from the site of initial wounding and loading, this method is less invasive than microinjection or biolistic bombardment. Tracers can even move up to the top of the plant, to the shoot apical meristem (SAM), a group of stem cells that gives rise to all the above-ground plant organs following germination.

For example, this approach reveals that PD in the SAM are dynamically regulated. During vegetative development, when the plant continuously produces new leaves, the cells at the SAM allow transport of small (~0.5 kDa, see below) symplastic tracers. However, during the transition from vegetative to reproductive development, when the plant starts producing flowers, PD at the apex are downregulated and no transport of tracers occurs (16, 17). Potentially, a signal molecule that regulates flowering is symplastically transported to the apex from leaves. The apex may then shut down further communication while it undergoes the profound morphological changes that accompany the switch to floral production. Interestingly, symplastic transport to the apex resumes once floral commitment is established. Such studies highlight the important role of PD during plant development.

The use of tissue-specific promoters to drive expression of fluorescent reporter proteins offered the next significant leap for PD research. In this approach, transgenic plants were constructed to express a soluble diffusible GFP (or GFP-fusion protein) in specific cells/tissues using a specific promoter. For example, soluble GFP expressed in the companion cells (CC) of source leaves (net export of photosynthetic products) of tobacco and Arabidopsis (13, 18) moves toward regions of new growth, such as sink leaves (net import of photosynthetic products) and newly-emerging floral organs. Strikingly, GFP was observed to move throughout all plant tissues and organs, albeit to more or less extents depending on the tissue. Thus the PD SEL is at least 27 kDa in many regions of the plant. Such movement implies that endogenous macromolecular signals may traffic the phloem to facilitate new development.

PD Function During Seedling Development

Phloem-loading together with novel fluorescent probes made it possible to track the cell-to-cell movement of symplastic probes, both locally and long distance, in whole seedlings just after germination from the seed (19). This approach is especially suited to the model plant Arabidopsis, as seedlings are small (~1 cm for the shoot and ~3 cm for the root of one-week-old seedling) and thus the whole plantlet can be viewed easily under the fluorescent microscope. Early studies used small (~0.5 kDa) tracers such as carboxyfluorescein (CF) diacetate and 8-hydroxypyrene-1,3,6-trisulfonic acid (HPTS). The ester (uncharged) form of CF diacetate freely moves across plasma membrane. A cytosolic esterase then converts this probe to the anionic membrane impermeable form, trapping CF in the cytoplasm. CF diacetate applied to cut leaves of Arabidopsis seedlings translocates *via* the phloem and unloads into growing root tips (19) to reveal symplastic coupling between young root cells (see Figure 2). HPTS was found to be a more reliable tracer as it is highly anionic and localizes entirely to the cytoplasmic compartment for intercellular transport *via* PD. HPTS loading revealed that the epidermis of the root becomes symplastically isolated from inner cells as development proceeds (20). While such phloem loading offers an excellent non-invasive means to monitor PD function, especially over long distances, this method can only measure the transport of small probes.

An elegant series of experiments revealed macromolecular movement in Arabidopsis seedling roots, and the precision whereby PD can control such movement. In particular, the SHORTROOT (SHR) transcription factor (TF) was found to move cell-to-cell in developing roots. The SHR TF is required for the normal differentiation of cortex/endodermal initial cells that control the formation of the endodermis. Surprisingly, transcription of *SHR* (shown by *in situ* mRNA localization and a transcriptional fusion of GFP to the promoter of *SHR*) is absent from cortex/endodermal initial cells or its daughter cells. Instead, *SHR* mRNA is present in the internally adjacent cells of the stele (Figure 3, inset). However, SHR protein (shown by immunolocalization and a translational fusion of GFP to the coding sequence of SHR) localizes to the stele and the cells of the adjacent cell layer of the endodermis, which includes the cortex/endodermal initial cells and the quiescent center (Figure 3). These results imply SHR-GFP traffics from the stele to a single adjacent layer of cells, where it functions to promote asymmetric cell division and endodermal cell fate.

Over the years, additional studies using genetics and *in situ* gene expression have revealed that other plant TFs can move cell-to-cell *via* PD. The classical example is the maize KNOTTED1 (KN1) protein, discovered in 1994 to move

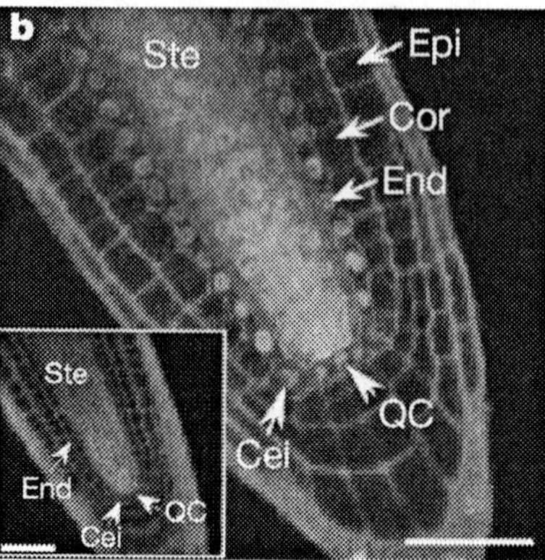

Figure 3. Intercellular movement of SHORTROOT (SHR) protein in Arabidopsis root. SHR proteins as GFP fusions (shown as grey regions) localize both in the stele (Ste) and endoderrmis (End), while *SHR* transcript locates only in the stele (inset). Reprinted from (59) with permission.

one cell layer in the shoot apex. KN1 is a homeodomain containing protein that regulates leaf and shoot meristem development. KN1 and its mRNA traffic from cell-to-cell in the shoot apical meristems and leaves (21). LEAFY (22) that controls floral meristems identity, and CAPRICE (23), that is central to root hair cell formation, are other examples of TFs that move cell-to-cell. More information about the movement of TFs *via* PD is reviewed in (9, 24-26).

For readers interested in additional studies on the role of PD during postembryonic development, we mention a few articles as starting points. Two recent studies analyze the transport of GFP tracers in early seedling leaves (27, 28) and roots (28). See also the role of symplastic communication in morphogenesis of postembryonic tissues such as gametophytes, leaf, root, stem, flower and shoot apical meristem of land plants and algae (20, 29-34). Note that symplastic isolation occurs in different manners and to various degrees, permanent *versus* transient, and complete PD closure *versus* reduced PD aperture, and symplastic domains differentiate into tissues with distinct structures and functions.

PD Function During Embryonic Stages: A Transient Assay to Identify PD Genes

While PD function and ultrastructure have been extensively analyzed, until recently few studies have addressed what genes control PD. Genetics is a powerful tool to isolate potential PD genes, yet PD research using genetics is quite limited (35-37). One obstacle to such an approach is, given that PDs are essential to plant growth, most PD mutants are unlikely to grow to adult plants. However, while PD mutants cannot be easily identified at the adult plant level, PD mutants should manifest early in development during embryogenesis. Such lethal PD mutants can be propagated as heterozygous plants that then display their homozygous defective phenotype in embryos segregating in seedpods or fruit. One Arabidopsis fruit (called silique) contains 40-60 seeds in which embryos are enclosed (think of a pea pod with 40 peas, but much smaller) (Figure 4A). Thus, 10-15 homozygous mutant embryos will be segregating in a single silique.

The next hurdle was to develop a strategy to test PD function during embryogenesis. First embryos need to be released from their seed coats. Seeds are extruded from siliques and collected in a glass slide. Application of a cover glass and slight pressure releases embryos. This extrusion process induces sublethal tears in plasma membranes and cell walls in the outermost cell layer of embryos. Such breaks provide initial entrance sites for symplastic tracers of various sizes (Figure 4D). Probes larger than PD SEL of the cells at the break site are trapped in the initial cells (Figure 4B,C) and cannot move, whereas tracers smaller than the PD SEL move cell-to-cell *via* PD (Figure 4D).

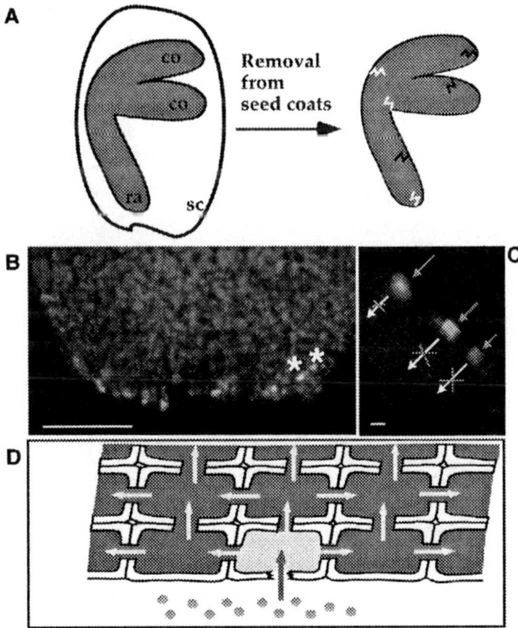

Figure 4. Uptake of symplastic probes in cells of Arabidopsis midtorpedo embryos. (A) When embryos are released from their seed coats, physical damage occurs in a subset of cells. As a result, small regions of cell walls and plasma membranes are broken to a sublethal level to provide an initial entrance site for uptake of symplastic tracers such as HPTS and F-dextran, which do not cross plasma membranes. Jagged lines indicate the most common site of damage. co, cotyledon; ra, radicle; sc, seed coat. (B) A small number of cells at the base of the detached cotyledons from midtorpedo embryos are cytoplasmically loaded with 10 kDa F-dextran (asterisks), yet further movement to neighboring cells does not occur. Scale bar, 50 μm. (C) A typical example of loaded cells in a region containing abrasion at the edge of the protodermal layer, marked as jagged lines in (A). Individual cells in the protodermal layer take up 10 kDa F-dextrans (arrows) and show cytoplasmic localization of the probe. However, subsequent movement of the probe is inhibited (arrows with X). Scale bar, 5 μm. (D) A diagram shows a partially broken cell wall and plasma membrane (jagged edge) may provide the initial entrance site for uptake of symplastic tracers, F-dextran or HPTS (circles). Further symplastic transport is then determined by the PD SEL and the size of symplastic tracers introduced. Reprinted with permission from (37).

In this transient assay, HPTS (0.5 kDa) or fluorescently (F)-labeled 10 kDa dextrans were exogenously introduced into developing embryos. HPTS moves through all cells of embryos throughout all stages of embryonic development examined (early heart to midtorpedo), demonstrating that the embryo is single symplast. However, the use of higher molecular weight tracers reveals that PD aperture is downregulated as development proceeds. 10 kDa F-dextrans are transported cell-to-cell in 50% of heart, 20% of early torpedo, and 0% of midtorpedo embryos. Thus, while symplastic connectivity remains (as measured by small tracers such as HPTS), PD SELs are altered during development.

Over 5,000 lines of Arabidopsis with an embryo defective phenotype were screened by the above assay to detect mutants that continued to traffic 10 kDa F-dextran at the midtorpedo stage. Fifteen lines, called *increased size exclusion limit of plasmodesmata* (*ise*), were identified (37). Two lines, *ise1* and *ise2*, are currently under investigation to identify their defective genes and characterize their role in PD function and/ or structure.

PD Function During Embryonic Stages: Analysis of Symplastic Domain Formation during Embryogenesis

Besides providing a genetic tool, embryos are innately interesting subjects for investigation of intercellular transport patterning. Embryogenesis is a critical stage of plant development that sets up basic body axes enabling the development of different tissues and organs. Arabidopsis embryos have regular pattern of cell divisions that allow the tracking of the origin of seedling structures back to specific groups of cells in the early embryo (38, 39). The seedling shows an apical–basal pattern along the main axis composed of structures such as shoot apical meristem (SAM), cotyledons, hypocotyl and root (Figure 5I). Clonal analyses and histological techniques predict the contribution of each embryonic cell to this body plan (40) (Figure 5I, compare heart and seedling). Generally, positional information determines the overall body pattern, and lineage-dependent cell fate specifies local patterning (40-42). Auxin signaling as well as differential gene expressions then facilitate specific morphogenesis (reviewed in (43, 44)).

Cell-to-cell signaling *via* PD is an important factor to coordinate embryonic development. However, until recently no studies have directly addressed PD function during embryogenesis. Now evidence suggests that PD also conveys positional information during axial patterning in late embryogenesis (see below). For these studies, stable (*versus* transient introduction of tracers) expression of GFP in specific regions of the embryo was investigated.

Subdomains Corresponding to Axial Body Pattern

Two different promoters were used to drive GFP expression in meristematic regions of Arabidopsis embryos. The *SHOOT MERISTEMLESS* (*STM*) promoter was used to express 1X, 2X and 3XsGFP (single 27 kDa, double 54 kDa and triple 81 kDa forms of sGFP) in the shoot apical meristem (SAM) and a subset of cells in the hypocotyl (45). In addition, the cell-type-specific enhancer of the J2341 line-induced expression of 2XsGFP in the SAM and the root apical

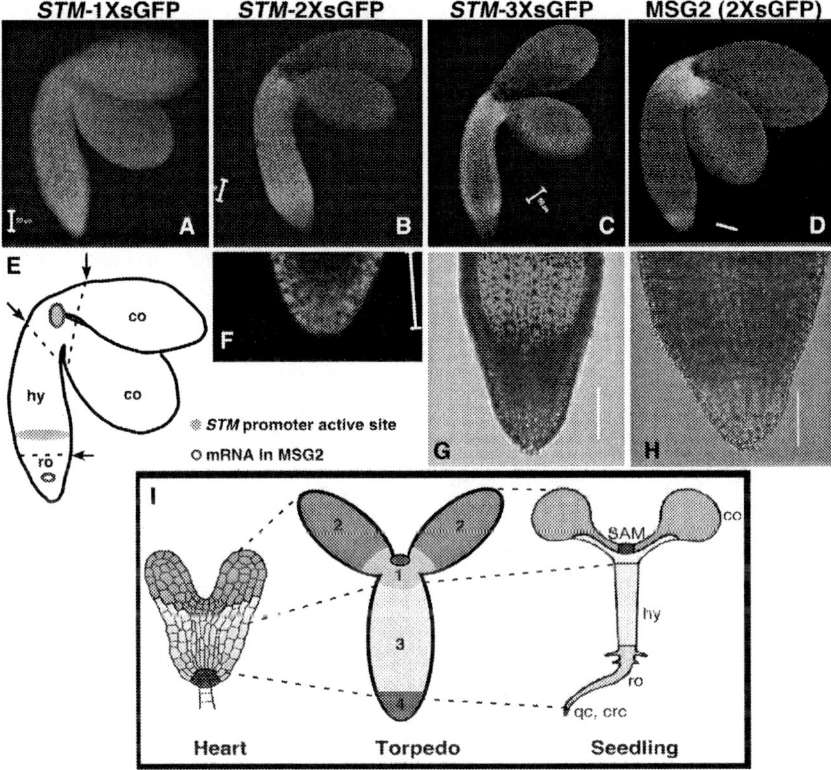

Figure 5. sGFP movement in Arabidopsis midtorpedo embryos. 1XsGFP expressed by the *STM* promoter in the SAM and the base of hypocotyls (hy) (E) freely moves throughout the whole embryo (A). 2XsGFP fails to move into cotyledons (co) (B) but moves to the root tip (F). 3XsGFP fails to move to the root (ro) as well as cotyledons (C,G). These results indicate the formation of at least two symplastic subdomains, e.g. the cotyledon and root. 2XsGFP expressed in the SAM and RAM in MSG2 line (E) stays within subdomains of the shoot apex and the root, respectively (D,H). These results, together with (B), reveal the boundary between the shoot apex and hypocotyl subdomains. Root subdomains from embryos in (C) and (D) are shown in larger magnification views under each whole midtorpedo image, and include quiescent center (qc), part of the RAM, and central root caps (crc). (E) Origin of MSG2-mediated expression is indicated by empty circles at SAM and RAM, and origin of *STM*-mediated expression is indicated by shaded circles at the SAM (same as MSG2) and the lower part of hypocotyl. (I) Four symplastic subdomains in torpedo embryos, shoot apex (1) including SAM (a dark circle), cotyledons (2), hypocotyl (3), and root (4) are extrapolated to the body parts in heart embryos and seedlings. Same shading in heart embryo and seedling represent regions of development with common clonal origins. Subdomains of the torpedo embryo, determined by their cell-to-cell transport *via* PD, also correspond to the apical-basal body pattern of the heart embryo (and seedling) by their positions; these regions are diagrammed with different shadings to indicate they were defined by a different assay. Scale bars, 50 μm. Reprinted with permission from (27).

meristem (RAM) in the MSG2 line (Figure 5E) (27). The subsequent movement of these various-sized tracers from their site of synthesis was monitored at three stages of embryogenesis to reveal two major findings. First, 2XsGFP (54 kDa) moves throughout the entire early heart embryo demonstrating that PD apertures (interconnecting cells to form a single symplast) in early embryos are quite dilated. Secondly, different regions of the embryo have distinct PD SELs defining symplastic subdomains by the midtorpedo stage. These subdomains correspond to the major regions of the apical-basal body axis, the shoot apex, cotyledons, hypocotyls and root. (See Figure 5 and legend) (27). These subdomains can be extrapolated to regions of the early embryo (and seedling) defined by gene expression profiles and clonal analyses (Figure 5I).

Boundaries Between Symplastic Subdomains of Cell-to-Cell Transport

The above data imply that there are boundaries between each of four symplastic subdomains where the embryo controls intercellular transport (45). Each boundary has a distinct PD SEL. For example, the boundary between the shoot apex and the cotyledons has a SEL between 27 and 54 kDa, as 1XsGFP but not 2XsGFP moves from the SAM to the cotyledons (Figure 5A, B, and E). The boundary between the hypocotyl and the root has a SEL between 54 and 81 kDa, as 2XsGFP but not 3XsGFP moves from the hypocotyl to the root (Figure 5F and G). The hypocotyl and shoot apex subdomains are indicated by the movement of 2XsGFP from its site of synthesis at the SAM and surrounding cells in MSG2, and its failure to move to the hypocotyl (Figure 5D and E). Movement of 2X and 3XsGFP in the hypocotyl subdomain results from upward movement from its site of synthesis (under the *STM* promoter) near the hypocotyl-root junction (Figure 5B, C, and E). The existence of the root and cotyledon subdomains was further investigated in transgenic plants expressing 1X or 2XsGFP fused to the P30 movement protein (MP) of *Tobacco mosaic virus* (TMV), also under the control of the same *STM* promoter (45).

TMV P30 localizes to PD in virus-infected cells (46) and in uninfected transgenic plants expressing P30 (47). TMV P30 acts as a molecular chaperone to bind the single-stranded viral RNA genome and targets this ribonucleoprotein complex to PD, where it triggers an increase in PD SEL (called gating) to facilitate movement of the TMV genome into adjacent uninfected cells (reviewed in (48)). In embryos, GFP-P30 targets to PD as in adult plants, and moves more extensively than similarly sized GFP tracers, confirming the functionality of P30. However, 1XGFP-P30 (57 kDa) and 2XGFP-P30 (84 kDa) behaved as the similarly-sized 2XsGFP (54 kDa) and 3XsGFP (81 kDa) in their inability to be transported into cotyledons and roots, respectively (45). These data reinforce the existence of boundaries between symplastic subdomains in embryos.

Further Refinement of Local Symplastic Subdomains

To date additional symplastic subdomains, corresponding to the protodermis and stele, have been observed. When 1XsGFP was expressed in the outermost protodermal layer of the hypocotyl, under the control of the Arabidopsis *GLABRA2 (AtGL2)* promoter, it moves uniformly inward to internal ground tissues and to neighboring protodermal cells in cotyledons at the heart stage (see Figure 3F of (49)). However, in the early torpedo stage, centripetal movement of 1XsGFP from the protodermis is reduced such that GFP signal intensity is now much weaker in ground tissues, while movement among cells in the protodermis continues (Figure 6A). Similarly, 1XsGFP expressed in the root tip, by the Arabidopsis *SUCROSE TRANSPORTER3 (AtSUC3)* promoter, freely moves to the hypocotyl in earlier stages (49), but becomes restricted to the stele in the midtorpedo stage (Figure 6B).

Note that the extent of symplastic movement is significantly affected by the location of the initial site of sGFP synthesis. 1XsGFP freely moves to every cell in embryos following expression in the SAM (27, 45), but its movement is limited to within the stele upon expression from a subset of cells in the root tip (49) (compare Figures 5A and 6B). It makes sense that PD in and around the SAM are more active than those in the root tip, as meristems are likely the source of morphological signals to enable patterning during embryogenesis. Future studies need to address how the SAM (and RAM) contribute to the formation of symplastic subdomains to determine the apical-basal body pattern, and how symplastic sudomains corresponding to various tissue types are controlled locally.

Symplastic Domains in Developing Seed Coats

The Arabidopsis seed coat consists of five cell layers, the innermost endothelial layer, followed by two cell layers each of inner and outer integuments. Two symplastic domains, corresponding to the outer and the inner integuments, were identified in developing seed coats (49). GFP expressed in the outer integument cannot move to the inner integument layers (Figure 6D). Similarly, GFP

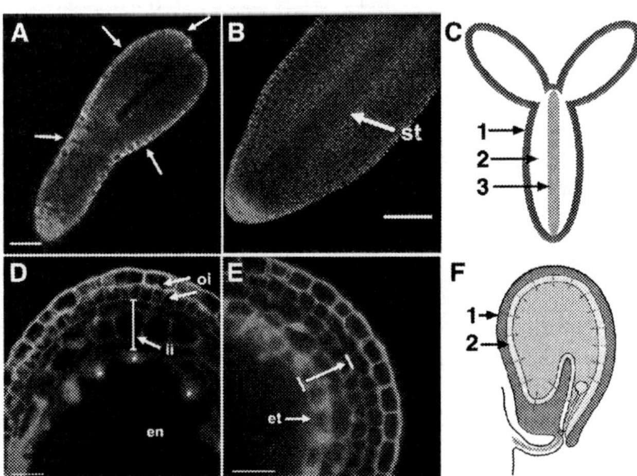

Figure 6. More subdomains in embryos and seed coats. The protodermis (A, arrows, and C1) and the stele (st) (B, C3) form subdomains where the movement of 1XsGFP, expressed by *AtGL2* and *AtSUC3* promoter, respectively, is allowed within domains but is reduced (A) or blocked (B) to cells beyond each domain. Outer integuments (oi) (D, F1) and inner integuments (ii) (E, F2) form separate symplastic domains where 1XsGFP movement is blocked across a boundary between the two domains. C2, ground tissues; en, endosperm; et, endothelium. Scale bars, 40 μm (A), 50 μm (B), 25 μm (D) and 20 μm (E). Reprinted with permission from (49).

expressed in the innermost endothelial layer moves to the inner integument layers, but cannot move to the outer integument layers (Figure 6E). Even small tracers such as HPTS (0.5 kDa) are not transported across the boundary between the outer and the inner integuments. Stadler *et al.* suggested that the outer integuments may provide a symplastic route for nutrient transport from maternal tissues to developing seeds, but that transfer between the outer integument and the inner integument, and the inner integument to the embryo may be apoplastic (49).

MORE APPROACHES TO IDENTIFY PD COMPONENTS

Although it is now established that PD have dynamic and critical roles in various aspects of plant life, no components specific to PD are known. In addition to the genetic approach mentioned above, several different approaches have been conducted in an effort to uncover structural or functional components of PD. A biochemical approach uncovered one *Nicotiana tabacum* NON-CELL AUTONOMOUS PATHWAY PROTEIN 1 (NtNCAPP1), from PD-enriched cell wall extracts as an interacting partner to a PD-trafficking protein (CmPP16) by affinity purification (50). NtNCAPP1 locates in the cell periphery and contains ER transmembrane domain which deletion blocks the movement of specific PD-trafficking proteins, suggesting that protein movement *via* PD is both selective and regulated. A plasmodesmal-associated protein kinase (PAPK) specifically interacts with plant viral proteins, such as TMV P30, and localizes to PD. Since P30 is known to manipulate PD, PAPK may act to regulate PD function (51).

A collection of random plant cDNA-GFP fusions and their localization in cells generated a library composed of GFP tags to specific plant organelles including PD (52). Another high-throughput screening where plant cDNA-GFP fusions were expressed by a viral expression system identified twelve proteins specifically localized to PD (53). A punctate pattern in cell walls is diagnostic for labeling and localization to PD. Half of the twelve-encoded proteins share no similarity with known proteins and may represent novel components of PD.

Proteomic technology is another approach to identify PD-specific proteins from purified PD or cell wall fractions enriched for PD (54). One protein found by several research groups is a class 1 reversibly glycosylated polypeptide (RGP). RGPs normally associate with the Golgi, but one RGP targeted to PD (55). The giant-celled green alga *Chara corallina* provides an advantageous system to apply proteomics (56) as cells are arranged in a single linear file and PD are localized to the cross walls between adjacent cells. Peptides isolated from PD-enriched cell wall fractions include previously known PD-associated proteins, validating the experiments, as well as novel proteins, providing new candidates for PD components.

PERSPECTIVES

The critical role of PD in plant development is supported by accumulating data of cell-to-cell movement of TFs critical in cell-fate determination. Recent data also suggest that RNAs, mRNAs and gene silencing RNAs (reviewed in (3, 57, 58)) also traffic *via* the vascular system and its connected PD. Besides identifying the cargo of PD, little is known about potential regulatory molecules that signal PD to allow selective movement of macromolecules. Furthermore, what are the exact mechanics of transport *via* PD? Diverse approaches including cellular, genetic and genomic tools will need to be synergistically applied to answer these questions.

ACKNOWLEDGMENTS

All current laboratory members performing research on PD are co-authors. PD research is supported by the National Institutes of Health, GM 45244.

REFERENCES

1 Zambryski, P. (2004) J. Cell Biol. 164, 165-168.
2 Oparka, K.J. (2004) Trends Plant Sci. 9, 33-41.
3 Lucas, W.J. and Lee, J. Y. (2004) Nat. Rev. Mol. Cell. Biol. 5, 712-726.
4 Heinlein, M. and Epel, B.L. (2004) Int. Rev. Cytol. 235, 93-164.
5 Roberts, A.G. (2005) in Plasmodesmata (K. J. Oparka, ed.) pp. 1-32, Blackwell.
6 Baluska, F., Cvrckova, F., Kendrick-Jones, J. and Volkmann, D. (2001) Plant Physiol. 126, 39-46.
7 Blackman, L.M., Harper, J. D. and Overall, R.L. (1999) Eur. J. Cell Biol. 78, 297-304.
8 Ding, B., Itaya, A. and Qi, Y. (2003) Curr. Opin. Plant Biol. 6, 596-602.
9 Kim, J. Y. (2005) Curr. Opin. Plant Biol. 8, 45-52.
10 Goodwin, P.B. (1983) Planta 157, 124-130.
11 Wolf, S., Deom, C.M., Beachy, R.N. and Lucas, W.J. (1989) Science 246, 377-379.
12 Lucas, W.J., Bouche-Pillon, S., Jackson, D.P., Nguyen, L., Baker, L., Ding, B. and Hake, S. (1995) Science 270, 1980-1983.
13 Oparka, K.J., Roberts, A.G., Boevink, P., Santa Cruz, S., Roberts, I., Pradel, K.S., Imlau, A., Kotlizky, G., Sauer, N. and Epel, B. (1999) Cell 97, 743-754.
14 Crawford, K.M. and Zambryski, P.C. (2000) Curr. Biol. 10, 1032-1040.
15 Crawford, K.M. and Zambryski, P.C. (2001) Plant Physiol. 125, 1802-1812.
16 Gisel, A., Barella, S., Hempel, F.D. and Zambryski, P.C. (1999) Development 126, 1879-1889.
17 Gisel, A., Hempel, F.D., Barella, S. and Zambryski, P. (2002) Proc. Nat. Acad. Sci. U.S.A. 99, 1713-1717.
18 Imlau, A., Truernit, E. and Sauer, N. (1999) Plant Cell 11, 309-322.
19 Oparka, K.J., Duckett, C.M., Prior, D.A.M. and Fisher, D.B. (1994) Plant J. 6, 759-766.
20 Duckett, C.M., Oparka, K.J., Prior, D.A.M., Dolan, L. and Roberts, K. (1994) Development 120, 3247-3255.
21 Kim, J.Y., Yuan, Z. and Jackson, D. (2003) Development 130, 4351-4362.
22 Wu, X., Dinneny, J.R., Crawford, K.M., Rhee, Y., Citovsky, V., Zambryski, P.C. and Weigel, D. (2003) Development 130, 3735-3745.

23 Wada, T., Kurata, T., Tominaga, R., Koshino-Kimura, Y., Tachibana, T., Goto, K., Marks, M.D., Shimura, Y. and Okada, K. (2002) Development 129, 5409-5419.
24 Hake, S. (2001) Trends Genet. 17, 2-3.
25 Wu, X., Weigel, D. and Wigge, P.A. (2002) Genes Dev. 16, 151-158.
26 Cilia, M.L. and Jackson, D. (2004) Curr. Opin. Cell Biol. 16, 500-506.
27 Kim, I., Cho, E., Crawford, K., Hempel, F.D. and Zambryski, P.C. (2005) Proc. Nat. Acad. Sci. USA 102, 2227-2231.
28 Stadler, R., Wright, K.M., Lauterbach, C., Amon, G., Gahrtz, M., Feuerstein, A., Oparka, K.J. and Sauer, N. (2005) Plant J. 41, 319-331.
29 Tilney, L.G., Cooke, T.J., Connelly, P.S. and Tilney, M.S. (1990) Development 110, 1209-1221.
30 van der Schoot, C. and van Bel, A. (1990) Planta 182, 9-21.
31 van der Schoot, C., Deitrich, M.A., Storms, M., Verbeke, J.A. and Lucas, W.J. (1995) Planta 195, 450-455.
32 Erwee, M.G. and Goodwin, P.B. (1985) Planta (Heidelberg) 163, 9-19.
33 van der Schoot, C. and Rinne, P. (1999) in Plasmodesmata: Structure, Function, Role in Cell Communication (A. van Bel and W. van Kesteren, eds.) pp. 225-242, Springer.
34 Kwiatkowska, M. (1999) in Plasmodesmata: Structure, Function, Role in Cell Communication (A. van Bel and W. van Kesteren, eds.) pp. 205-224, Springer.
35 Russin, W.A., Evert, R.F., Vanderveer, P.J., Sharkey, T.D. and Briggs, S.P. (1996) Plant Cell 8, 645-658.
36 Provencher, L.M., Miao, L., Sinha, N. and Lucas, W.J. (2001) Plant Cell 13, 1127-1141.
37 Kim, I., Hempel, F.D., Sha, K., Pfluger, J. and Zambryski, P.C. (2002) Development 129, 1261-1272.
38 Mansfield, S.G. and Briarty, L.G. (1991) Can. J. Bot. 69, 461-476.
39 Jurgens, G. and Mayer, U. (1994) in a Colour Atlas of Developing Embryos (J. Bard, ed.) pp. 7-21, Wolfe.
40 Scheres, B., Wolkenfelt, H., Willemsen, V., Terlouw, M., Lawson, E., Dean, C. and Weisbeek, P. (1994) Development 120, 2475-2487.
41 Poethig, R., Coe, E. and Johri, M. (1986) Dev. Biol. 117, 392-404.
42 Saulsberry, A., Martin, P.R., O'Brien, T., Sieburth, L.E. and Pickett, F.B. (2002) Development 129, 3403-3410.
43 Berleth, T. and Chatfield, S. (2002) in The Arabidopsis Book (C. Somerville and E. Meyerowitz, eds.) pp. 1-22, ASPB.
44 Laux, T., Wurschum, T. and Breuninger, H. (2004) Plant Cell 16 Suppl, S190-202.
45 Kim, I., Kobayashi, K., Cho, E. and Zambryski, P.C. (2005) Proc. Nat. Acad. Sci. U.S.A. 102, 11945-11950.
46 Tomenius, K., Clapham, D. and Meshi, T. (1987) Virology 160, 363-371.
47 Atkins, D., Hull, R., Wells, B., Roberts, K., Moore, P. and Beachy, R.N. (1991) J. Gen. Virol. 72, 209-211.
48 Ghoshroy, S., Lartey, R., Sheng, J. and Citovsky, V. (1997) Annu. Rev. Plant Physiol. Plant Mol. Biol. 48, 27-50.

49 Stadler, R., Lauterbach, C. and Sauer, N. (2005) Plant Physiol. 139, 701-712.
50 Lee, J.Y., Yoo, B.C., Rojas, M.R., Gomez-Ospina, N., Staehelin, L.A. and Lucas, W.J. (2003) Science 299, 392-396.
51 Lee, J.Y., Taoka, K.I., Yoo, B.C., Ben-Nissan, G., Kim, D.J. and Lucas, W.J. (2005) Plant Cell 17, 2817-2831.
52 Cutler, S.R., Ehrhardt, D.W., Griffitts, J.S. and Somerville, C.R. (2000) Proc. Nat. Acad. Sci. USA 97, 3718-3723.
53 Escobar, N.M., Haupt, S., Thow, G., Boevink, P. Chapman, S., Oparka, K. (2003) The Plant Cell 15, 1507-1523.
54 Faulkner, C., Brandom, J., Maule, A. and Oparka, K. (2005) Plant Physiol. 137, 607-610.
55 Sagi, G., Katz, A., Guenoune-Gelbart, D. and Epel, B.L. (2005) Plant Cell 17, 1788-1800.
56 Faulkner, C.R., Blackman, L.M., Cordwell, S.J. and Overall, R.L. (2005) Proteomics 5, 2866-2875.
57 Baulcombe, D. (2002) Curr. Biol. 12, R82-84.
58 Voinnet, O. (2002) Curr. Opin. Plant Biol. 5, 444-451.
59 Nakajima, K., Sena, G., Nawy, T. and Benfey, P.N. (2001) Nature 413, 307-311.

ROOT-KNOT AND CYST NEMATODE PARASITISM GENES: THE MOLECULAR BASIS OF PLANT PARASITISM

Thomas J. Baum[1], Richard S. Hussey[2] and Eric L. Davis[3]

[1]Department of Plant Pathology
Iowa State University
351 Bessey Hall
Ames, IA 50011
[2]Department of Plant Pathology
University of Georgia
[3]Department of Plant Pathology
North Carolina State University

INTRODUCTION

Roundworms of the Nematoda comprise one of the largest animal phyla on Earth (1). They inhabit diverse terrestrial and aquatic niches through adaptations of a spectrum of trophic groups, including parasites that threaten human, animal and crop plant health. The most well-known nematode, *Caenorhabditis elegans*, is a native soil-dwelling microbivore that has emerged as a premier model for animal biology and genomics (2). While studies of *C. elegans* provide a blueprint of fundamental nematode biology, recent advances in molecular genetics of parasitic nematodes indicate specific divergence in adaptations of nematodes for obligate parasitism of an array of plant and animal host species (3-8). Identifications of the molecular tools enabling a particular mode of parasitism by nematodes are providing some intriguing discoveries about the nature of parasite evolution.

Genetic Engineering, Volume 28, Edited by J. K. Setlow

Plant-parasitism by nematodes can be distinguished by which plant part is parasitized and the length of time the nematode feeds from a plant cell. Plant-parasitic nematodes, or phytonematodes, are considerably larger than a host plant cell, so a single, unmodified plant cell cannot sustain nematode feeding throughout the parasite's life cycle. This critical host–parasite balance is manifested in the adaptation of two very different phytonematode groups: migratory parasites that feed while moving from plant cell to cell and sedentary parasites that first modify plant cells in order to be able to continuously feed in one location as their bodies enlarge and they become immobile. These two groups frequently are likened to primitive *versus* highly evolved forms of parasitism, respectively. While this nomenclature probably does not describe the two modes of parasitism adequately since all parasitism more than likely encompasses highly-evolved traits, this distinction serves well in describing the different levels of complexity of the two different parasitic modes. Emphasis is placed in this review on adaptations of sedentary phytonematodes that induce dramatic changes in host feeding cells to sustain parasitism in one location.

MAJOR SEDENTARY PHYTONEMATODES

The root-knot nematodes, *Meloidogyne* spp. and the cyst nematodes, *Heterodera* and *Globodera* spp., are sedentary parasites of roots of many crop plant species that collectively incite billions of dollars in annual crop losses around the world. While both nematode groups use very similar parasitic strategies to complete their life cycles (Figure 1), they employ different mechanisms to carry out their strategies. In each group, the motile juvenile molts to the second-stage (J2) and hatches from the egg in soil. The infective J2 follows environmental and host cues in soil to locate tissues near the plant root tip that it will penetrate. Infective juveniles of root-knot nematodes and cyst nematodes differ somewhat in their means of migration and apparent preference for feeding location near the vascular tissue of host plant roots, which shall not be revisited here (9). More substantial differences become obvious once feeding commences. If initiation of feeding is successful, the sedentary parasitic phase ensues, leading to nematode growth and three subsequent molts to the reproductive adult stage. Both root-knot nematodes and cyst nematodes transform initial feeding cells into elaborate feeding sites that share a dense cytoplasm, altered cell walls, duplication of their genetic material and increased metabolic activity. However, root-knot nematode and cyst nematode feeding sites differ in ontogeny and appearance. The root-knot nematode induces substantial enlargement and changes in a small group of initial feeding cells around the nematode head and turns each of them into a discreet "giant-cell" from which the nematode feeds in sequence (Figure 2A). In each giant-cell, the nucleus undergoes repeated divisions resulting in a multinucleate state. A cyst nematode, on the other hand, induces changes in a single initial feeding cell, which then are reciprocated in neighboring cells, including cells that are not necessarily in direct contact with the nematode. These changes culminate in the fusion of many modified cells, sometimes involving over 200 cells, to form one large multinucleate cytoplasm called a syncytium (Figure 2B). Nuclei of syncytial cells undergo endoreduplication of their DNA content but do

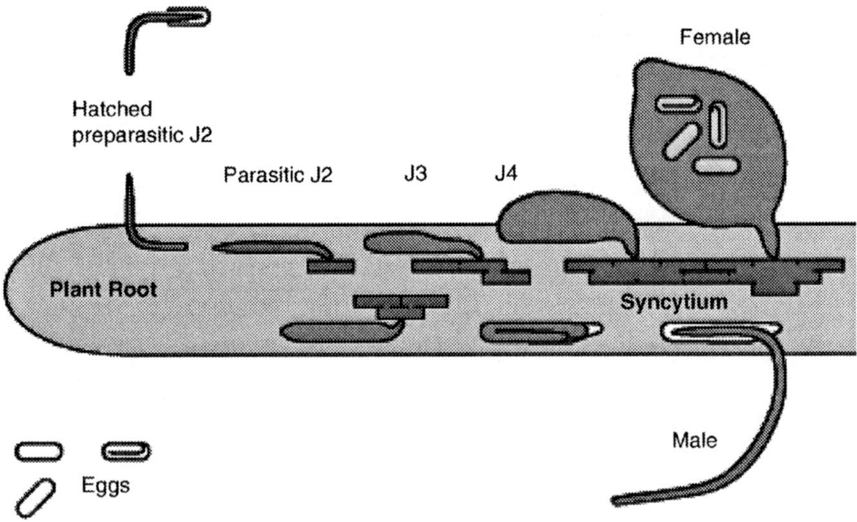

Figure 1. Cyst nematode life cycle. Second-stage juvenile (J2) cyst nematodes hatch from eggs in the soil and become parasitic by penetrating into the root of a host plant. Close to the root vascular tissue, parasitic J2 become sedentary and induce the formation of feeding sites called syncytia, which consist of fused root cells. Feeding from its syncytium, a nematode enlarges and matures through the third (J3) and fourth juvenile (J4) stage into either an adult female or a male nematode. Males regain mobility and exit the plant root to fertilize the still sedentary females (Drawing by J. de Boer).

not divide. The elaborate changes in morphology of both syncytia and giant-cells are accompanied by dramatic alteration in gene expression in the affected plant cells (10). Interestingly, root-knot nematodes and cyst nematodes in general also differ in the fact that most root-knot nematode species have broad host ranges whereas cyst nematodes have much smaller groups of host plants. A current hypothesis is that both nematodes use different strategies to induce their respective feeding sites and that giant-cell induction by the root-knot nematode targets a plant mechanism that is widely conserved among plant species, thereby allowing parasitism of many host plants. On the contrary, for the formation of syncytia, cyst nematodes may target molecular plant mechanisms that are divergent among different plants, and, therefore, individual cyst nematode taxa can only infect relatively small groups of plants.

ADAPTATIONS FOR PLANT PARASITISM

Plant-parasitism is thought to have evolved at least three times independently (3), but morphological adaptations for plant parasitism are surprisingly similar among all plant-parasitic nematodes. Most notably, all plant-parasitic nematodes are equipped with a stylet (hollow mouth spear) to pierce cell walls and allow solute exchange between plant and parasite. Furthermore, plant-parasitic nematodes have well-developed secretory gland cells associated with their esophagus that produce secretions released through the stylet into host

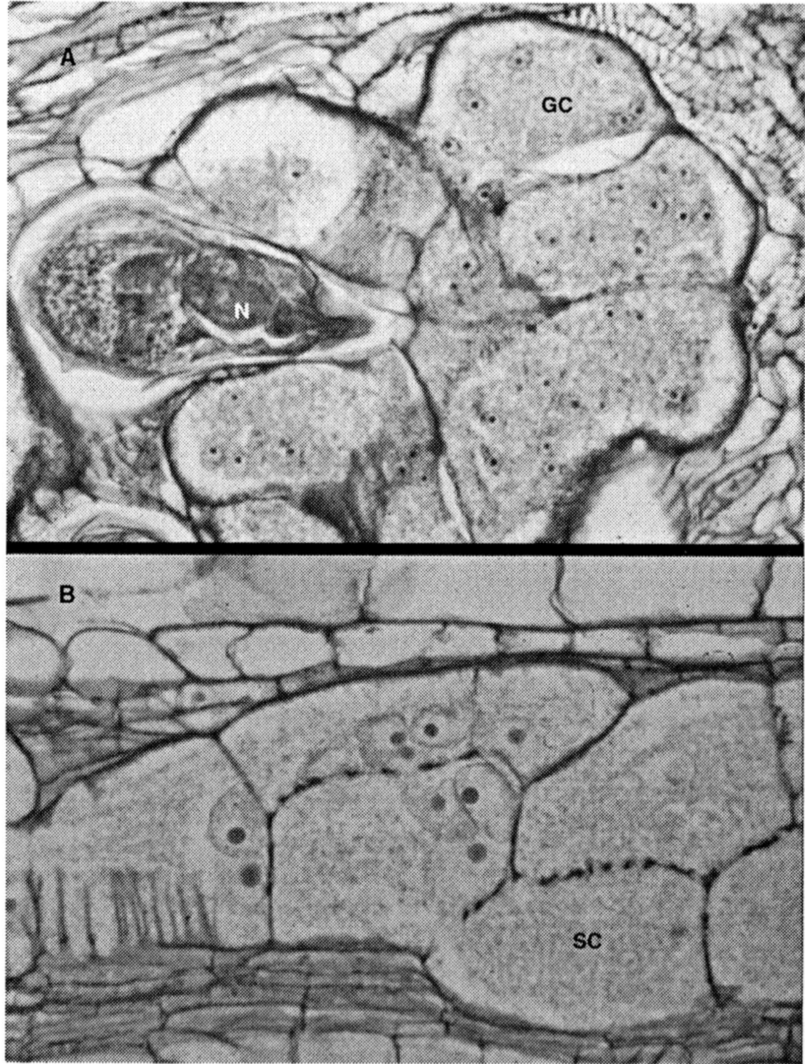

Figure 2. Feeding sites of root-knot and cyst nematodes. (A) Root-knot nematodes (N) induce the formation of giant-cells (GC) in the roots of their host plants. Each giant-cell contains multiple nuclei, which are visible in this figure (unknown source). (B) Cyst nematodes induce the formation of syncytia by fusion of individual syncytial cells (SC) through cell wall dissolution. Perforated cell wall remnants are clearly visible in this panel (Pictures by B. Endo).

tissues. Interestingly, the development of enlarged secretory cells associated with the esophagus also exists in nematode parasites of animals but is notably absent from microbivorous nematodes like *C. elegans* (7). In the case of the root-knot nematodes and cyst nematodes, as is the case with the other tylenchid phytonematodes, there are three esophageal glands, one dorsal and two subventral glands (Figure 3). Even though these structures are called glands, they are de-facto single

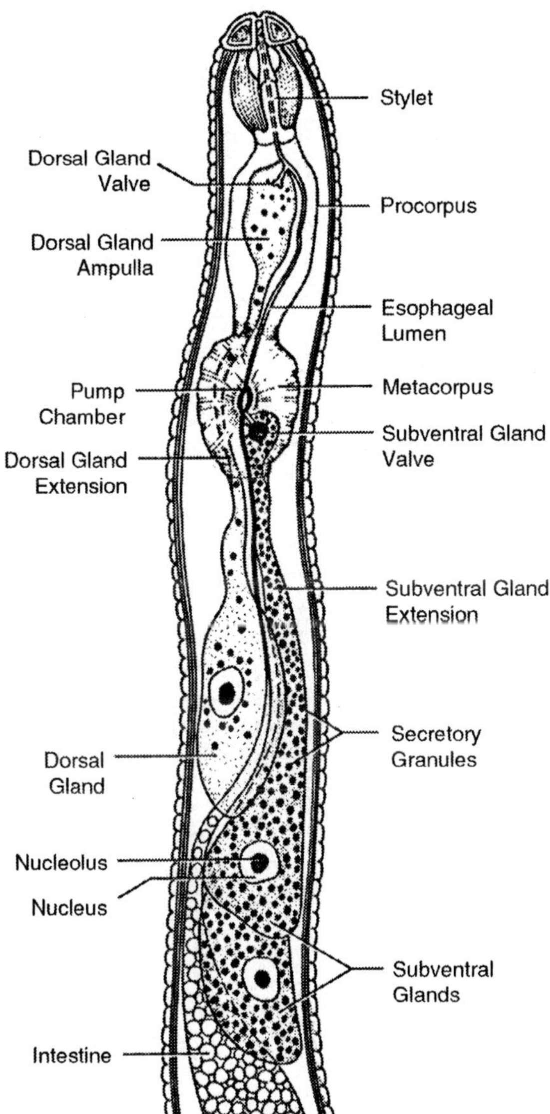

Figure 3. Anterior end of a second-stage juvenile cyst nematode. The anterior end of cyst nematodes harbors major adaptations for plant parasitism, particularly the stylet and the three esophageal glands (one dorsal gland and two subventral glands). This anatomy is completely shared by the root-knot nematodes (Drawing by R. Hussey).

cells, each having long cytoplasmic extensions that are connected through valves to the lumen of the esophagus (11). Secretory proteins are synthesized in these cells and packaged into membrane-bounded secretory granules. The granules move anteriorly through the gland extensions, and their contents are released into the esophageal lumen by exocytosis *via* the respective gland-cell valve. While in root-knot and cyst nematodes the two subventral gland cell extensions open into the esophageal lumen immediately posterior to a muscular pump chamber in the median bulb, the dorsal gland cell extends anterior in the esophageal wall to empty through a valve into the esophageal lumen at the base of the stylet. This morphological difference implies different functions of the requisite glands, and this assumption is confirmed by the dramatically-different developmental appearance of the gland cells during the developmental cycle of the root-knot and cyst nematodes. As early as in fully-developed J2 in the egg, the extensions of the subventral glands of root-knot and cyst nematodes are packed with secretory granules whereas the dorsal gland extension is relatively empty. During the transition from host-root penetration to feeding site induction and maintenance, the subventral glands become smaller and less active while the dorsal gland enlarges and increases in activity for the remainder of the parasitic cycle. The movement of contents from both esophageal gland cell types for secretion through the stylet has been documented in elegant video-enhanced microscopy of plant-parasitic nematodes within roots (12-15). While there was initial conviction that only the dorsal gland, due to the opening of its cytoplasmic extension near the base of the stylet, has a function in parasitism, the subventral glands were thought to function only in secreting digestive proteins destined for the nematode intestine. This restricted role of the subventral glands has now been convincingly refuted, as will be discussed later on. The developmental changes in gland cell activity (and secretory proteins noted below) during different stages of parasitism, and the conduit to the parasitized host cell through the stylet, point to secretions from both gland types as direct adaptations to promote parasitism.

NEMATODE PARASITISM GENES AND THEIR PRODUCTS

Plant-parasitic nematodes are parasites that become pathogens only secondly, depending on the human perception of the severity of parasitism, i.e., whether the parasitism causes visible, economically-damaging symptoms. Therefore, the molecular mechanisms allowing a nematode to infect a plant are those mechanisms making a nematode a parasite—and not a pathogen. Hence, the genetic determinants that enable a nematode to infect plants are appropriately named parasitism genes. It is obviously of utmost interest to determine what makes a nematode a plant parasite, i.e., to determine which nematode genes are responsible for the ability to parasitize plants. In the widest sense, genes underlying morphological adaptations (e.g., the stylet), behaviors (e.g., host-finding or mating), or abilities (reproductive or survival strategies) that promote a successful parasitic lifestyle represent essential and often specific adaptations for parasitism. However, this global view, while academically interesting, does not focus on the direct molecular interactions between parasite and host, which are at the biochemical basis of plant-parasitism by nematodes. A more focused view of nematode parasitism genes targets those genes that code for proteins released from the nematode that directly interact with host molecules to promote the parasitic interaction. For reasons cited above, genes encoding secretions produced by nematode esophageal gland cells are prime candidates as nematode parasitism genes (Figure 4). Studies have confirmed that nematode stylet secretions

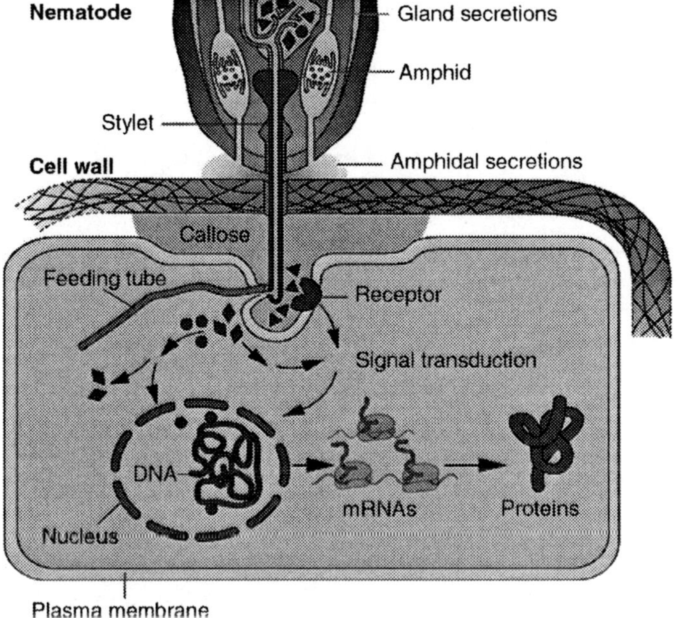

Figure 4. Parasitism gene functions. Parasitism genes in a narrow sense code for secretory proteins directly involved in the nematode-plant interaction. These parasitism proteins are secreted through the stylet into the parasitized plant where they have important functions during the induction of feeding cells like giant-cells and syncytia. Parasitism proteins may function as extracellular or intracellular ligands or signal transduction components, be imported into the nucleus, or act on cytoplasmic components, all of which could modify the recipient plant cell. Furthermore, parasitism proteins have functions during feeding like the formation of feeding tubes (Drawing by R. Hussey).

produced in the esophageal gland-cells are proteinaceous and not nucleic acids (16), suggesting that secretions are translated directly from parasitism gene transcripts. Molecules released or secreted from other nematode body regions could also be involved in parasitism, either as encoded proteins or as the products of metabolic pathways. There are examples of candidate parasitism proteins produced in the amphids (chemosensory organs found at the head of nematodes) or even the hypodermis (the inner living cell layer of the nematode's body wall). The best studied examples of parasitism proteins are those produced in the esophageal glands and released as secretory proteins. These proteins are synthesized as preproteins with N-terminal signal peptides that target the nascent protein chain during translation of the parasitism gene mRNA to the endoplasmic reticulum. There, the signal peptide is cleaved off and the mature protein passes along the secretory pathway. However, there are a few examples of parasitism protein candidates that presumably use a different mode of secretion not requiring a signal peptide. Nonetheless, even when considering such exceptions, the majority of currently known parasitism genes are expressed exclusively in the esophageal-glands and code for secretory parasitism proteins requiring a signal peptide.

POTENTIAL ROLES OF PARASITISM PROTEINS

When considering parasitism genes in the narrow sense described above, i.e., esophageal-gland-expressed genes coding for secretory proteins released through the nematode stylet, an array of possible involvements of parasitism proteins in the nematode life cycles can be postulated. First of all, nematodes need to penetrate the roots of their host plants and migrate through root tissues. Considering the moderate size of root-knot nematodes and cyst nematode infective J2s, cell walls pose formidable obstacles, and, as will be discussed later on, both nematodes use a mixture of cell-wall-digesting enzymes to break structural integrity of plant cell-walls. In addition to these important functions, the most impressive achievement appears to be the nematode-directed formation of the elaborate feeding cells by root-knot nematodes (giant-cells) and cyst nematodes (syncytia). As mentioned above, the nematodes need to communicate with mostly differentiated root cells and induce the development of the parasitized cells into the different feeding cell types. Furthermore, the nematodes need to maintain these cells, which probably include suppressing plant defenses and/or cell death programs that may be activated during parasitism. Finally, video footage of a feeding cyst nematode (12, 15) and micrographs of other nematode feeding sites including root-knot nematodes (17-19), clearly show following the release of secretions through the dorsal-gland-valve the formation of a tubular structure (feeding tube) at the stylet orifice inside the cytoplasm of the feeding cell (Figure 4). Hence, feeding tube formation along with feeding cell maintenance during food uptake, more than likely, are roles of parasitism proteins. The size exclusion of molecules ingested by root-knot nematodes and cyst nematodes has been documented to be between 28 and 40 kD (20, 21), suggesting that the feeding tube acts as a molecular sieve.

PARASITISM GENE IDENTIFICATION

The identities of parasitism proteins have intrigued scientists, and an array of approaches to identify parasitism genes and proteins have been devised and tried. Most of these approaches targeted the esophageal-glands because of their obvious involvement in parasitism. Antibodies specific to esophageal-gland antigens were generated using *in vitro* purified nematode stylet secretions or fractions of nematode homogenates and used to screen cDNA expression libraries (22) or to affinity purify the nematode antigens (23). Furthermore, efforts were expended to directly identify purified stylet-secreted proteins (24-29). Also, the mining of ever-growing databases containing the nucleotide sequences of expressed genes (Expressed Sequence Tags, ESTs), revealed parasitism gene candidates because of their similarity to already identified parasitism genes from other nematode species or to proteins with obvious functions in parasitism (30-35). Finally, gene expression at the RNA level at different time points or in different nematode tissues was assessed in hopes of identifying parasitism genes because of their developmental expression patterns or their localized expression in the esophageal-glands (36-42). However, the most exhaustive and direct approach to identify parasitism genes targeted the esophageal-glands directly *via* microaspiration of gland-cell cytoplasm followed by the construction and mining of gland-specific cDNA libraries (37, 38, 43-49). All these approaches have been detailed and compared in recent reviews and will not be repeated here (4-6, 8). One of the greatest conceptual advances in nematology over the last decade has been the discovery that sedentary plant-parasitic nematodes produce in their esophageal-glands a large array of secretory proteins with putative functions in parasitism (5). Determining the identity of parasitism genes, however, is only the first step toward unraveling the mechanisms of plant parasitism by nematodes. Understanding the functions of the parasitism proteins, individually or in concert, currently represents the biggest obstacle in this research.

KNOWN PARASITISM GENES

A current list of root-knot and cyst nematode genes with known putative functions in parasitism, mostly based on similarities to characterized proteins in other organisms, is presented in a recent review (8) and shall not be repeated here. In addition to the parasitism proteins with similarity to characterized proteins, there is an even larger number of parasitism genes from root-knot and cyst nematodes for which no similarities to characterized proteins in other organisms exist (44, 47).

It is an interesting observation that when parasitism proteins are similar to known proteins, this similarity usually is not with proteins from *C. elegans*, a non-parasitic nematode whose genome is fully sequenced. Rather, if similarities to nematode proteins are found, these similarities are frequently only with proteins from other parasitic nematodes. Most frequently, however, similarities are with proteins from bacteria, fungi, or plants for which there are no functions in nematodes. For example, plant-parasitic nematodes produce cellulases and pectinases, yet there are no substrates for these enzymes found within the nematode. Similarly, these nematodes do not have a shikimate pathway, yet they produce a key enzyme of this pathway. Also, nematode parasitism proteins sometimes represent secretory versions of known cellular effector proteins. These curiosities all point in one direction, namely that these nematode proteins do not have a function within the nematode but function as instruments of parasitism when secreted within the parasitized plant.

CURRENT HYPOTHESES OF PARASITISM PROTEIN FUNCTIONS

Despite the fact that the majority of parasitism protein candidates currently known are without similarity to characterized proteins, interesting conclusions can be drawn, nonetheless, from a relatively small group of parasitism proteins. In this group, similarities of parasitism proteins with functionally characterized proteins from other organisms and the functional characterization of parasitism proteins that already has been accomplished allow the formulation of credible working hypotheses about mechanisms of parasitism used by root-knot nematodes and/or cyst nematodes.

Cell-Wall-Digesting Enzymes

As already mentioned above, it has been established that root-knot nematodes and cyst nematodes use a mixture of enzymes to soften root-cell-walls, which should aid in penetration through the root epidermis as well as migration within root tissues. To date, there have been cellulase and pectinase genes described for root-knot nematode (38, 47, 48, 50) and cyst nematode species (23, 44, 46, 51-53). Discovery of cellulase genes in the soybean and potato cyst nematodes represented the first major breakthrough in parasitism gene discovery and was of particular interest because at that time, no cellulase genes had been reported from animals (23). In addition, nematode cellulases were highly similar to bacterial proteins, which raised the interesting hypothesis that a certain subset of parasitism genes was actually acquired by horizontal gene transfer (23, 52). Similarly, pectinases had not been reported from animals as well, and the nematode pectinase proteins were of the pectate lyase type found in fungi and bacteria, cyst and root-knot nematodes; (35, 47, 48, 51, 54) or the polygalacturonase type of bacteria (root-knot nematode; 55). An involvement of these enzymes in penetration and migration is backed by the fact that cell-wall-digesting enzymes are produced and secreted during nematode penetration and migration and to a much smaller extent, or not at all, during the later sedentary stages (48, 50, 56-58; A. Elling and T. J. Baum, unpublished data). Interestingly, males of cyst nematodes, who regain mobility and leave host roots, reinitiate cellulase production during this life stage (56, 58). Very convincing support is also gained from experiments in which genes for cell-wall-digesting enzymes are inactivated by gene-silencing techniques (see below) and J2 infectivity is reduced (59). While it is clear that these enzymes are used for the purpose of cell-wall softening, it is not clear why the nematodes have large gene families for some of these proteins and what exactly are the functions of the individual gene family members (57). Similarly, the function of cellulose-binding proteins discovered in root-knot and cyst nematodes remains elusive (36, 44, 45, 47): do these proteins function in concert with cellulase enzymes that lack a cellulose-binding domain or do these proteins have functions in their own right? The latter is suggested by the finding that in planta overexpression of a bacterial cellulose-binding domain led to accelerated cell growth (60). Research outside the realm of sedentary nematodes also reported beta-1,4-endoglucanase genes from the lesion nematode *Pratylenchus penetrans* (61), which is a migratory parasite that obviously also requires successful means to breach plant cell-walls. Very interestingly, a cellulase of the beta-1,3-endoglucanase type recently was reported from the pinewood nematode *Bursaphelenchus xenophilus* (a fungus-feeding, insect-vectored nematode living in pine trees) where it is hypothesized of being involved in nematode feeding from fungal mycelium (62).

Expansins

In addition to the ability to break down covalent bonds found in plant cell-walls (i.e., through cellulases and pectinases) there is evidence that the potato cyst nematode also secretes a protein having the ability to break non-covalent bonds. This activity is accomplished by an expansin-like protein discovered in the potato cyst nematode (41), which represented the first confirmed report of such a protein outside the plant kingdom. Expansins soften cell-walls by breaking non-covalent bonds between cell-wall-fibrils, thereby allowing a sliding of fibrils past each other. The resultant plant cell-wall softening could be demonstrated for the potato cyst nematode expansin parasitism protein (41). No such genes have been found in root-knot nematodes or other cyst nematodes to date.

Metabolic Enzymes

Discoveries in both root-knot (39, 49) and cyst nematodes (32, 44) identified parasitism genes coding for proteins similar to chorismate mutases. These enzymes catalyze the conversion of the shikimate pathway product chorismate to prephenate. This process represents a key regulatory mechanism determining the ratio of the aromatic amino acids phenylalanine and tyrosine on one hand and tryptophan on the other. Consequently, this regulatory activity influences the production of the metabolites that have these amino acids as precursors, among which auxin and salicylic acid are of particular interest in plant-parasite interactions. The plant shikimate pathway is found in the plastids from where chorismate also is translocated to the plant cytoplasm. According to the current understanding of chorismate mutase function, nematode-secreted chorismate mutases will deplete the cytoplasmic chorismate pool leading to an increased translocation of chorismate from the plastids, effectively decreasing synthesis of plastid-produced chorismate-dependent metabolites like auxin or salicylic acid. Expression of a root-knot nematode chorismate mutase gene in soybean hairy roots produced an auxin-deficient phenotype, which gave rise to this model of chorismate mutase function (63). A lack of salicylic acid production in response to nematode chorismate mutase injection could result in a downregulation of plant defenses. In line with a putative function in defense deactivation, it was observed that chorismate mutases represent a polymorphic gene family in soybean cyst nematodes and that presence and expression of certain gene family members correlates with the nematodes' ability to infect certain soybean genotypes harboring soybean cyst nematode resistance genes (64, 65).

Ubiquitination/Proteasome Functions

Targeted and timed protein degradation is a final and powerful means to regulate gene expression. Cyst nematodes apparently are using this mechanism to alter gene expression in parasitized plant cells since these nematodes appear to secrete proteins involved in polyubiquitination, i.e., the process that specifically decorates proteins with ubiquitin protein molecules thereby targeting these proteins for degradation. This hypothesis is founded in the discovery that cyst nematodes produce secretory isotypes of otherwise purely cytoplasmic proteins involved in the ubiquitination pathway, namely ubiquitin itself, along with proteins (i.e., RING-Zn-Finger-like and Skp1-like proteins) similar to those found in the E3 ubiquitin protein ligase complex (42, 44). An additional level of complexity exists in the fact that the nematode-produced ubiquitin molecules also contain a short C-terminal extension with unknown function. Unlike known non-nematode ubiquitin extension proteins (66), the nematode extension apparently is not a ribosomal protein and, therefore, its function remains unknown.

Small Bioactive Peptides

Recent scientific progress has begun to establish significant roles for small peptides in plant development (67). For example, the small extracellular ligand CLAVATA3 in Arabidopsis has been established as a key factor determining shoot meristem differentiation (68). It was particularly intriguing when it was discovered that the soybean cyst nematode produces a small parasitism peptide with a conserved C-terminal motif found in CLAVATA3-like ligand peptides (46, 69). Expressing the cDNA of this soybean cyst nematode CLAVATA3-like peptide in the *clavata3* (*clv3*) Arabidopsis mutant restored the wild-type phenotype, thereby confirming a first case of ligand mimicry in phytonematology (70). In other words, the soybean cyst nematode has evolved a secreted ligand for an endogenous plant receptor in order to parasitize the host plant successfully. Functionality also has been shown for a small 13 amino acid root-knot nematode parasitism peptide that previously had been discovered (47). This root-knot nematode peptide, when produced in planta, increased the rate of cell division in root meristems and was shown to bind to a plant transcription factor of the SCARECROW family (G. Huang and R. S. Hussey, unpublished data). This finding represents a first discovery of a direct regulatory interaction between nematode and plant proteins and, therefore, represents a powerful starting point for further exploration of this pathosystem. Considering the established importance of small peptides in signaling roles in plant development as well as plant–parasite interactions, it also will be interesting to determine if the small C-terminal extension of the cyst nematode ubiquitin extension proteins mentioned above (42, 44) will have regulatory functions in the recipient plant cell. Additional support for a role of small peptides in nematode-plant interactions is presented by an unknown peptide fraction smaller than 3 kDa isolated from potato cyst nematode secretions. This protein fraction was shown to have biological activity by stimulating proliferation of tobacco leaf protoplasts and human peripheral blood mononuclear cells (71).

Nuclear Localized Parasitism Proteins

Analyses of parasitism proteins using computational approaches to predict protein localization and fate identified a significant subset of putative parasitism proteins with predicted nuclear localization signals (NLS), i.e., protein domains that mediate active uptake into the nucleus (44, 47). However, these proteins also contained N-terminal signal peptides directing them into the endoplasmic reticulum. This conflict can be resolved by postulating that NLS-containing nematode parasitism proteins first are targeted to the nematode gland-cell endoplasmic reticulum and the secretory pathway and only after secretion into a plant cell are they taken up into the plant nucleus. In testing this hypothesis, the active uptake of nematode parasitism proteins into plant nuclei has been shown for a small group of cyst nematode parasitism proteins (42; A. Elling and T. J. Baum, unpublished data). It will be of utmost interest now to decipher protein functions within the plant nucleus.

RanBPM

In a project comparing gene expression patterns among discrete developmental stages of the potato cyst nematode, a group of parasitism gene candidates was identified (40). Further analyses of these genes revealed the presence of a small family of genes coding for secretory proteins with high similarity to proteins binding to the small G-protein Ran, so-called RanBPMs (Ran-Binding Protein in the Microtubule organizing center). Several of these genes were expressed in the dorsal-gland (72). Exact functions of RanBPMs remain elusive and appear to be complex and diverse including the regulation of the cell cycle. Therefore, it is a tempting hypothesis that potato cyst nematode proteins with similarity to RanBPM may have a function in regulating the cell cycle activities observed in developing syncytia (72). As a first step it remains to be seen if Ran-binding activity or an effect on plant cell phenotype can be demonstrated for these nematode peptides.

Venom-Allergen Proteins

The parasitism proteins listed above are similar to functionally characterized proteins from other organisms, which allowed the formulation of clearly-defined hypotheses about protein function during parasitism. On the other hand, there are those parasitism protein candidates that are similar to known proteins whose functions, however, are still unknown or too diverse. This intriguing group of parasitism proteins contains representatives from root-knot nematodes (73) and cyst nematodes (37, 44) that are collectively called venom-allergen proteins (vaps). Gene sequences for these venom proteins were first described from hymenopteran insects (74), and vaps were also identified as secreted proteins (ASP) in the animal-parasitic nematode *Ancylostoma caninum* (75). Genes encoding vaps have since been found in other nematodes, including parasites as well as the free-living *C. elegans*. While several of these proteins were found to be secreted, or in the case of soybean cyst nematodes to be expressed in the subventral-glands (37), their function remains elusive.

Calcirecticulin

In a similar development, a calcirecticulin-like protein preceded by a signal peptide was identified as being produced in the subventral-glands of a root-knot nematode (27). Calcirecticulin-like proteins are secreted from other parasitic nematodes and, therefore, are good candidates for being involved in parasite-host interactions (76, 77). However, the puzzling array of putative or demonstrated calcirecticulin functions reported (76) make it difficult to postulate a function in plant parasitism by root-knot nematodes.

Annexin

Similarly, the mRNA for a secretory isoform of an annexin-like protein was identified as being expressed in the dorsal-gland of the soybean cyst nematode (44).

Annexin genes represent a large family coding for calcium-dependent phospholipid-binding proteins with a wide range of reported functions. Therefore, no clear postulation about annexin functions in cyst nematode parasitism can be made at this time. An annexin gene also had been identified from the potato cyst nematode *G. pallida*. This gene coded for a protein that was immunodetected in the excretory/secretory products of this nematode despite the fact that the protein did not contain a signal peptide and was not present in the esophageal-glands (78).

Chitinase

Also, there is an example of a parasitism protein with clearly defined function but no obvious role for this function at the time of the protein production. This putative parasitism protein is a chitinase identified in the subventral glands of the soybean cyst nematode (43). The only report of chitin in a nematode has been in the egg shell (79) and chitinases have been discussed as having a role in nematode hatch. However, *in situ* expression analyses (43) as well as microarray expression analyses (A. Elling and T. J. Baum, unpublished data) clearly demonstrate that this chitinase gene is not expressed in the eggs but that it shows a strong expression peak during the early phases of parasitism after penetration. As with many other parasitism proteins, further research has to explore a role for chitinase production during this stage of parasitism.

PARASITISM GENES IN A WIDER SENSE

In addition to the aforementioned parasitism proteins that satisfy the requirements of being produced exclusively in the esophageal-glands and harboring an N-terminal signal peptide, a small number of potentially interesting candidate genes that differ in at least one of these criteria have been identified.

Peroxidase

It appears likely that nematodes deploy means to cope with reactive oxygen species (ROS) produced by the host plant as a defense means in response to nematode attack (80). Such ROS-detoxifying enzymes have been reported in the form of peroxidases from the potato cyst nematode (33, 81). Peroxidase genes are expressed in the potato cyst nematode hypodermis and the peroxidase proteins accumulate on the nematode body surface presumably to detoxify ROS.

FAR

Another example of secreted nematode proteins with potential roles in negating plant defenses is a surface associated retinol- and fatty acid-binding (FAR) protein found in the potato cyst nematode *G. pallida*. This protein was found to bind to lipids that are precursors of the jasmonic acid signaling pathway as well as plant defense compounds (82). The reported accumulation of this protein at the nematode surface makes it a strong candidate for a protein that could interfere with plant defense mechanisms.

SXP/RAL-2

Another hypodermis-expressed gene coding for a secretory protein as well as a related gene expressed in glands associated with the anterior chemosensory organs (amphids) were identified from the potato cyst nematode *G. rostochiensis*. Both genes code for proteins of the nematode SXP/RAL-2 family, for which no functions could be ascertained to date (31).

Avr Protein

The amphids of a root-knot nematode were found to express a secreted protein that, while of unknown primary function, appears to represent a nematode avirulence protein, i.e., a protein whose presence leads to the initiation of effective plant resistance mechanisms triggered by the tomato *Mi* resistance gene (83). It will be of utmost interest to decipher the primary role of this protein and the mode by which it appears to trigger a resistance response.

14-3-3

A final protein with the potential of being involved in nematode parasitism has been discovered in the root-knot nematode *M. incognita*. This dorsal-gland-expressed gene codes for a protein of the 14-3-3 family that appears to be secreted despite lacking an N-terminal signal peptide (28). 14-3-3 proteins are well conserved in eukaryotes with a diverse spectrum of putative functions, and a role in nematode parasitism, if any, remains obscure.

DIFFERENCES BETWEEN ROOT-KNOT NEMATODES AND CYST NEMATODES

As mentioned above, root-knot and cyst nematodes use similar strategies to enable their sedentary parasitic life styles. However, it appears that these nematodes use very different tools of fulfilling their strategies because root-knot nematodes usually have wide host ranges and cyst nematodes narrow ones and the ontogeny of their feeding sites (giant-cells *versus* syncytia) is very different in certain aspects. Fully characterizing the root-knot nematode and cyst nematode parasitism genes should provide more definite answers. When assessing the currently identified panels of parasitism genes found in root-knot nematodes and cyst nematodes, one can find support for the hypothesis that root-knot nematodes and cyst nematodes use different molecular tools for their otherwise similar life habits—at least during the sedentary phase of parasitism. During the migratory phase, both nematode groups (root-knot nematodes and cyst nematodes) use cellulase and pectinase enzymes produced in their subventral glands in order to penetrate into and migrate through plant roots. Also, during the early phases of parasitism both nematode groups produce cellulose-binding proteins and venom-allergen proteins for unknown reasons and both root-knot nematodes and cyst nematodes produce chorismate mutase enzymes potentially to inactivate plant host defenses. However, as a first significant difference, a cyst nematode was shown to use an expansin parasitism protein to soften host cell walls, which is a group of proteins so far not identified in root-knot nematodes. Even more profound differences exist beyond these early stages of parasitism. While the soybean cyst nematode uses a small ligand with similarity to CLAVATA3-like proteins and appears to employ an ubiquitination pathway, no such proteins were discovered in root-knot nematodes. Instead, a large percentage of parasitism protein candidates without any database similarities (including cyst nematode genes) were found in root-knot nematodes. Also, while both root-knot nematodes and cyst nematodes produce a high proportion of unknown parasitism protein candidates in their dorsal-gland, root-knot nematodes appear to produce a relatively large proportion of unknown parasitism proteins also in their subventral glands. Of course, these assessments can only rely on the current state of knowledge and can only be completely validated when all parasitism proteins of several species of both nematode groups are identified. In summary, parasitism protein identities so far confirm that root-knot nematodes and cyst nematodes share certain aspects of their parasitic strategies but that key components of their arsenals of molecular tools likely are very different.

WHICH GLAND HAS WHICH FUNCTION?

Over the years, theories about the functions of the subventral glands *versus* the dorsal-gland have changed considerably. Early observations led to the conclusion that only the dorsal-gland is involved in direct parasitism functions because the subventral-glands emptied into the esophagus behind to the pump chamber, suggesting a transport of subventral gland-produced proteins only posteriorly into the intestine. However, the first parasitism gene to be identified was a subventral-gland-expressed gene coding for a cyst nematode cellulase which was definitively secreted through the nematode stylet, thereby, refuting the earlier hypotheses about subventral-gland proteins (84). When more and more cell wall-digesting or cell wall-modifying enzymes that were produced in the subventral-glands and secreted through the stylet were identified, it was plausible to speculate that the subventral-glands function during migration whereas the dorsal-gland proteins would be involved in mechanisms needed for feeding site formation and feeding. This was even more intriguing when considering that many of the subventral-gland-produced parasitism proteins were candidates for horizontal gene transfer acquisition by plant-parasitic nematodes because these proteins were most similar to prokaryotic or fungal proteins or had not even been reported from animals. In other words, it seemed intriguing to think of the subventral-gland as expressing a group of parasitism genes with a narrow function during nematode migration and obtained from other organisms. However, soon exceptions were reported that showed subventral-gland-produced proteins without known function during migration and that were produced even after the nematode had become sedentary. Currently, it appears most likely that subventral-gland-produced proteins have a pronounced but not exclusive role during nematode migration. Apparently, subventral-gland and dorsal-gland function in concert during the induction phases of feeding sites. Only the later stages, when feeding site maintenance and feeding appear to be the main functions, seem to be the dorsal-gland's exclusive domain.

FUNCTIONAL CHARACTERIZATION OF PARASITISM PROTEINS

The above-mentioned hypotheses about parasitism protein function have been formulated because of similarities of parasitism proteins with known, already characterized proteins. Additionally a variety of approaches have been employed to advance parasitism protein functional characterization. Such approaches are particularly important when considering that the majority (>70%) of currently identified parasitism proteins have no similarity to known proteins, particularly, those parasitism proteins produced in the dorsal-gland. A panel of molecular approaches is currently being used that will be instrumental in advancing knowledge of parasitism protein function. The following paragraphs provide short summaries of some of the most powerful approaches currently used.

Parasitism Gene Expression Profiling

Determining the exact locale of gene expression is of utmost interest for any gene-of-interest and of particular importance for parasitism genes. Expression in the 'wrong' cell can eliminate a gene from consideration while the opposite can provide the needed confirmation. Case in point, specific expression in one or more esophageal-glands has been a key criterion for parasitism gene discovery. Techniques for assessing gene expression at the mRNA as well as the protein level have been well established in the form of *in situ* mRNA hybridization (85) as well as *in situ* immunofluorescence analyses (86) (Figure 5). Similarly, not only the location but also the timing of expression is extremely valuable since it can provide insight into gene function. Characterization of cellulase parasitism genes, for example, was advanced when the developmental expression of cellulase gene family members was assessed, which determined a likely involvement of cellulases in the migratory phases of nematode parasitism – an observation complementing the fact that cellulases most likely aid in digesting cell walls during penetration and migration (56, 58). Analysis of gene expression over time can be accomplished using the *in situ* methods mentioned above, although processing high numbers of gene candidates proved to be challenging (44, 47, 56, 58). An alternative for the temporal assessment at the mRNA level is presented by microarray analyses. This approach has been employed with glass slides containing a small set of soybean cyst nematode cDNA sequences (87) as well as with Affymetrix GeneChips® containing oligonucleotide probe sets for more than 7,000 soybean cyst nematode genes. This latter approach identified the temporal expression of all currently known parasitism gene candidates along with all currently known soybean cyst nematode genes from eggs to adult female (A. Elling and T.J. Baum, unpublished data).

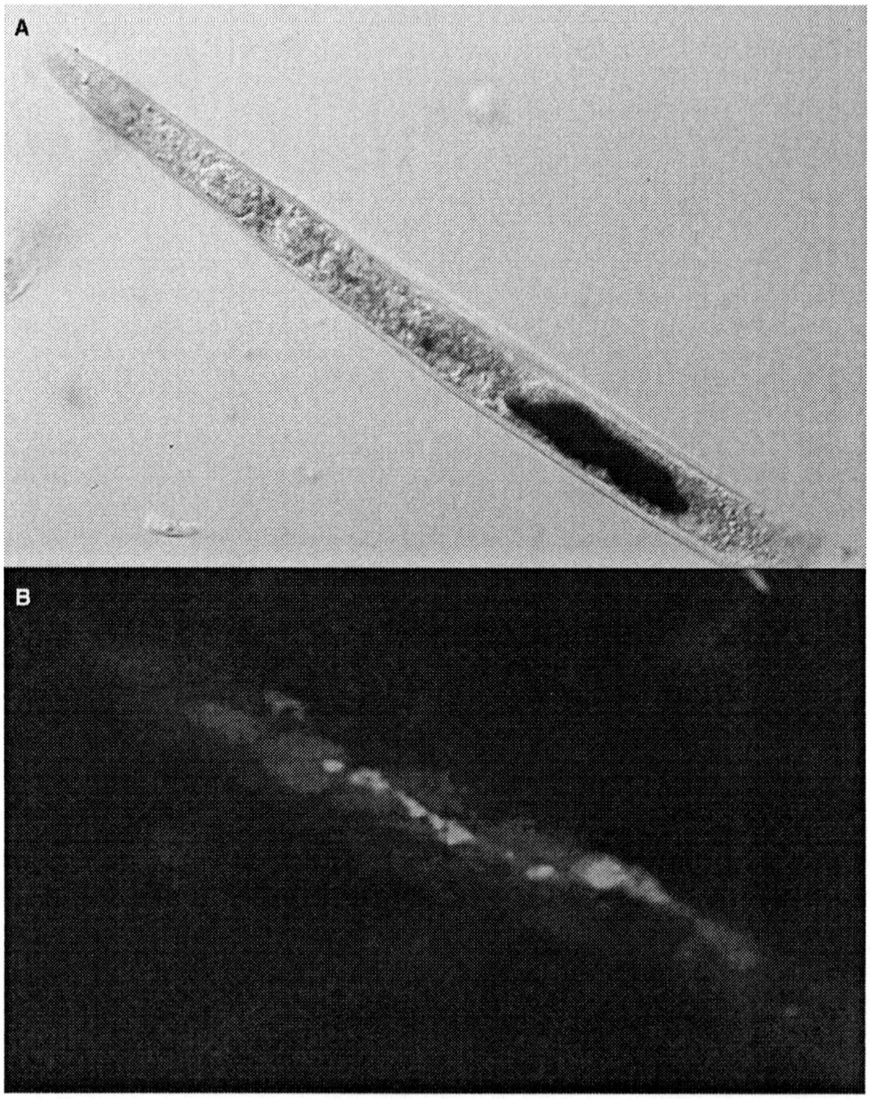

Figure 5. Assessing parasitism gene expression in the nematode. Determining the locale of gene expression in the nematode can be accomplished on the mRNA as well as the protein level. *In situ* hybridization (A) reveals the mRNA accumulation of this parasitism gene in the subventral esophageal-glands (dark stain). This result is confirmed by immunolocalization (B) of the corresponding parasitism protein in the same nematode glands (green fluorescence) (Pictures by G. Huang).

In Planta Localization of Parasitism Proteins

An equally crucial area of research is the documentation of secretion of nematode parasitism proteins inside the plant tissue. This not only again provides meaningful insight into protein function but it represents the ultimate proof that a protein-of-interest in fact can serve a direct function in nematode–plant interactions. Unfortunately, documenting a secreted protein is challenging at best, as several major hurdles pose obstacles to achieving success. A most elegant approach would be the production of a protein-of-interest as a reporter protein fusion in the nematode itself to follow the protein's fate when secreted into plant tissue/cells. Unfortunately, to date, no reliable protocols for the transformation of plant-parasitic nematodes have been published. The only other alternative is immuno-detection in planta, which requires high quality antibodies. But even with a specific antibody or serum, detection of a nematode protein in planta is difficult and frequently inconclusive. So far, documentation of in planta accumulation of parasitism proteins (Figure 6) has been very limited (54, 63, 84). Problems arise from the small amount of protein secreted from nematodes and the fact that once deposited into a plant cell, nematode proteins most likely form complexes with plant proteins or are processed, both of which can seriously impede antibody binding. Additionally, these obstacles don't even take into account the low probability of fixing a plant specimen and preparing an appropriate tissue section of the exact place and time when a given parasitism protein is secreted. A large number of sera to soybean cyst nematode and root-knot nematode parasitism proteins has been produced recently (E.L. Davis, R.S. Hussey and T.J. Baum, unpublished data) and these challenging assays are under way.

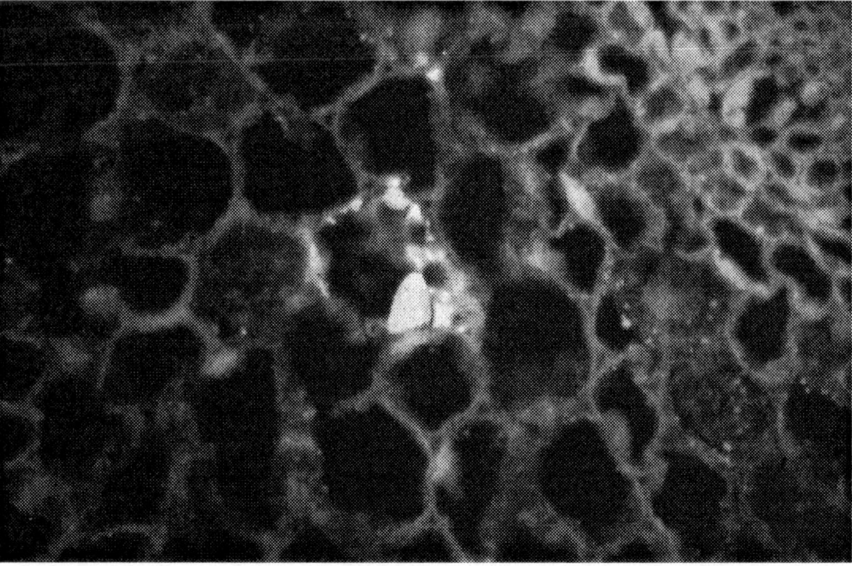

Figure 6. In planta accumulation of parasitism proteins. This section shows the head of a cyst nematode second-stage juvenile that was migrating through a soybean root. Immunolocalization of a cellulase parasitism protein clearly shows the accumulation of this parasitism protein (green fluorescence) along the migration path and on the outside of the nematode cuticle, thereby confirming in planta secretion of this protein (Picture by X. Wang).

Intracellular Localization of Parasitism Proteins

Another useful approach is the assessment of the subcellular localization of a parasitism protein once delivered to plant cells. Although not a substitute for the in planta localization, it represents a good tool for further characterization. For this purpose, nematode parasitism proteins are produced in planta as fusion proteins with reporter proteins like GUS or gfp. A significant number of cyst nematode parasitism proteins has been shown to be transported into the plant nucleus using this approach (42; A. Elling and T.J. Baum, unpublished data) (Figure 7).

Plant Expression of Parasitism Genes

Expression of parasitism genes in planta can be used to establish the fate of the encoded protein as well as to assess phenotypic changes of the plant or

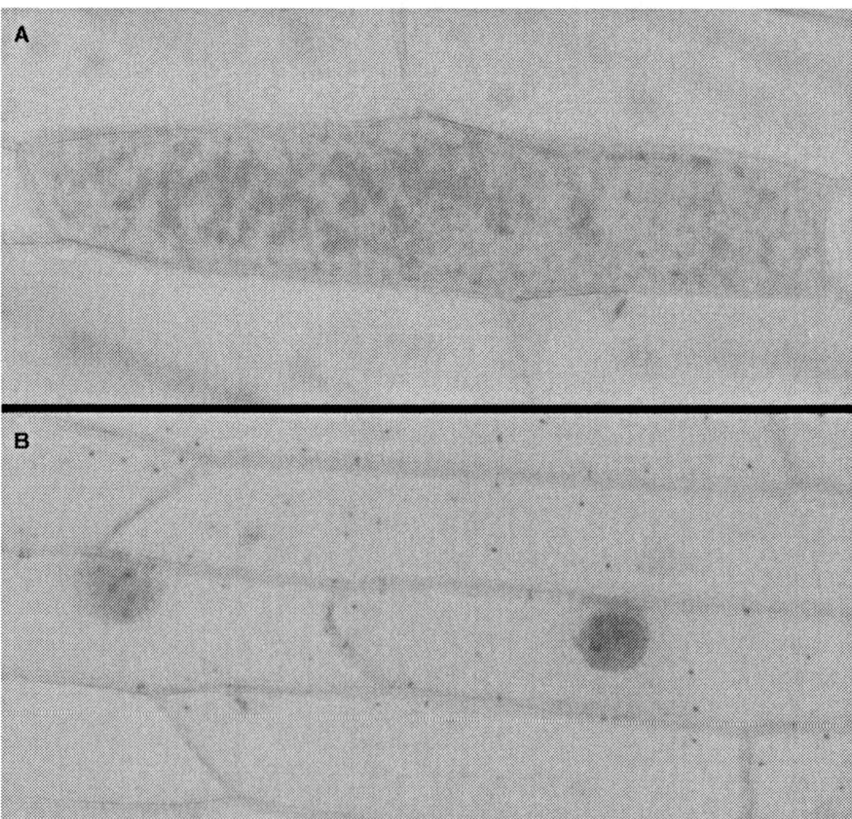

Figure 7. Intracellular localization of parasitism proteins. Translational fusion of parasitism proteins with the GUS reporter gene allows the visualization of protein localization. (A) A protein without specific targeting domains accumulates in the cytoplasm of onion epidermal cells. (B) A parasitism protein containing a nuclear localization signal is efficiently transported into the onion cell nucleus and accumulates there exclusively (Pictures by A. Elling).

parts thereof resulting from its overexpression. Because root-knot nematode and cyst nematode parasitism is accompanied by dramatic plant changes, it can be speculated that individual parasitism proteins will contribute to these changes by changing a certain aspect of the normal plant phenotype. Of particular interest here is the decision whether to include the parasitism protein signal peptide, i.e., whether one suspects the parasitism protein to function in the plant cell cytoplasm or in the apoplast. Furthermore, the choice of promoter is crucial and can influence the results and the conclusions to be drawn from a particular experiment, as this choice determines in which tissues, when, and to what strength a given parasitism gene is transcriptionally turned on. Of particular interest here are inducible promoters that can be used to customize parasitism gene expression. Expression of a few parasitism genes so far resulted in detectable phenotype changes in wild-type plants (G. Huang and R.S. Hussey, unpublished data). Particularly interesting could also be the expression of a parasitism gene suspected to code for an avirulence protein in a resistant host background because correct parasitism gene expression should trigger a visible resistance response.

Mutant Complementation

Another very powerful application of parasitism gene expression in heterologous organisms is the use of mutants as recipient organisms with the goal to restore the wild-type phenotype, thereby proving parasitism protein function. This approach was used to determine chorismate mutase function by complementing a bacterial chorismate mutase mutant (39, 49). As already mentioned above, producing the soybean cyst nematode parasitism protein containing a CLAVATA3-like conserved domain in the Arabidopsis *clv3* mutant restored the wild-type phenotype. Unfortunately, it is rather the exception that suitable protein similarities exist and well-defined mutants are available. Nonetheless, when successful, such complementation data provide strong support for a protein function.

Gene Silencing

Reverse genetics have been powerful in many biological systems because understanding gene functions can be achieved by inactivating a gene-of-interest. With the recent increased understanding of double-stranded (ds) RNA-induced gene silencing pathways, so-called RNA interference (RNAi), reverse genetics also became available to plant-parasitic nematodes despite our inability to stably transform these organisms. The obstacle remains how to expose plant-parasitic nematodes to the RNA species required to induce the RNAi mechanism. The observation that RNAi can be initiated in *C. elegans* by ingestion of dsRNA molecules (88) provided an important breakthrough for plant-parasitic nematodes. Incubating plant-parasitic nematodes in solutions containing dsRNA complementary to regions of a gene-of-interest led to a decrease of that gene's mRNA abundance (59, 89-91). In some cases phenotypes could be associated with this mRNA decrease, thereby revealing valuable insights into putative gene functions. For example, inactivating cellulase genes in the potato cyst nematode *G. rostochiensis* by soaking in dsRNA resulted in a decrease in nematode parasitism (59). Undoubtedly, further use of this technique will be a crucial advancement in determining contributions to and roles in parasitism of individual parasitism genes. A variation to this approach is currently explored in which dsRNA is produced in planta (92) within the nematode-induced feeding sites with the goal to allow a direct uptake of siRNA (<28 kD) by the feeding nematode. In addition to revealing parasitism protein function using this approach, the identification of which parasitism genes are essential for plant parasitism could lead to the development of novel and durable resistant transgenic plants using the RNAi technology.

Search for Interacting Proteins

It is likely that many parasitism proteins once delivered into the host plant will engage in interactions with plant proteins. Knowing the identity of such plant proteins has the potential to advance the understanding of parasitism protein function plus it will open additional avenues for further research. For example, parasitism proteins translocated into the plant nucleus will have to interact with plant cytoplasmic proteins to enable nuclear uptake where they in turn may interact with other proteins in order to exert their main function. Promising approaches to identify plant proteins that interact with nematode proteins are yeast-two-hybrid analyses and direct identification of such proteins through affinity purification. However, such approaches are not straight-forward and prone to many artifacts. To make matters worse, the confirmation of a suspected protein-protein interaction is equally tricky. Conceptual problems exist for example when considering that nematodes appear to secrete multiple protein and that it is conceivable that more than one nematode parasitism protein needs to be present to accomplish correct binding to plant proteins. Also, nematode parasitism proteins pass through the nematode gland-cell secretory pathway and there could be subject to modifications like glycosylation and/or cross-linking, which could alter protein-protein interactions. None of these protein modifications is easily reproduced in standard assays targeting the identification of interacting proteins. Nonetheless, first successes have been reported. As already mentioned above, SCARECROW transcription factor-like proteins were found to interact with a small parasitism peptide from a root-knot nematode, which could be confirmed through co-immunoprecipitation (G. Huang and R.S. Hussey, unpublished data).

PRESENT AND FUTURE

When assessing the putative identities of the parasitism genes described above, one can find four groups of parasitism gene similarities. In the first group, nematode parasitism gene candidates are found that have similarity to non-nematode, non-animal, or even non-eukaryotic genes. Classic examples are the nematode cellulase genes that code for proteins very similar to bacterial cellulases. Such genes are strong candidates for genes acquired by horizontal gene transfer. In a second group, one can find nematode parasitism genes that are similar to genes found in other, non-parasitic nematodes. For example, the annexin or venom-allergen genes mentioned above. These genes are found throughout the

Nematoda and other animals and may have evolved in root-knot nematodes and cyst nematodes to allow their protein products to assume functions during parasitism. In a third class, there are genes without similarity among animal genes but whose protein products exhibit weak similarities with plant proteins or domains thereof, and which apparently can function in the context of plant regulatory mechanisms. Good examples are the small soybean cyst nematode protein with similarity to the plant CLAVATA3 ligand or the root-knot nematode peptide that binds to a plant SCARECROW transcription factor. The final group of parasitism genes is the largest and contains genes coding for secretory proteins with unknown identity. These parasitism genes present the most difficult candidates to investigate for function. A combination of increased genomic data, bioinformatics and *in vivo* functional analyses discussed above, particularly RNAi, will be critical to unravel potential roles of these "pioneer" proteins in parasitism.

This review is a snapshot of our current understanding and thinking regarding the molecular basis of nematode parasitism of plants. The next decade holds tremendous promise in advancing our knowledge of parasitism genes and proteins and it will be interesting to compare our knowledge now with the level attained then. It will be particularly interesting to learn more about the reports that plant-parasitic nematodes release cytokinine plant hormones (93) or that root-knot nematodes potentially use a NOD factor-like signaling compound (94). These discoveries open the door to an additional realm of complexity and difficulty, namely the fact that nematodes may release compounds other than proteins in order to determine the outcome of their interactions with plants. It will be particularly rewarding to determine the origins of such compounds, which genes are involved in their synthesis, and their potential functions in parasitism.

As the genome sequencing efforts of the first species of phytonematodes are just beginning, the existing cache of expressed parasitic nematode genes underscores the urgency for robust analyses of gene function in the post-genomic era. The obligate nature of nematodes as parasites makes application of "routine" *C. elegans* technologies challenging, yet recent success with applications of RNAi to parasitic nematodes are encouraging. These emerging technologies not only provide critical analyses of gene function, but they offer the exciting potential to identify novel targets to interfere in parasite biology to protect human, animal, and crop health.

Finally, it is for the most part completely unclear how the nematodes manage secretion of their parasitism proteins. That is, are parasitism proteins secreted as mixes within individual secretory granules in the esophageal-glands or are they separately packaged? Is the secretion of individual proteins a process under the regulation other than gene expression, i.e., can the nematode deliberately secrete one protein and not another while both are present within the same gland? These are just a few of the truly interesting biological questions that need to be answered to obtain a more complete picture of plant-parasitic nematode parasitism. However, already now, available knowledge opens the way to several avenues to create novel means for nematode control, which is the main charge that warrants research on plant-parasitic nematodes in the first place, and maybe the most interesting developments of the near future will be the realization of new mechanisms to render plants resistant to plant-parasitic nematode attack.

REFERENCES

1 Poinar G. (1983) A Natural History of Nematodes. Prentice-Hall, Englewood Cliffs, NJ.

2 Ainscough, R., Bardill, S., Barlow, K. et al. *C. elegans* Sequencing Consortium (1998) Science 282, 2012-2018.

3 Blaxter, M., De Ley, P., Garey, J., Liu, L.X., Scheldeman, P., Vierstraete, A., Vanfleteren, J., Mackey, L., Dorris, M., Frisse, L., Vida, J. and Thomas, W. (1998) Nature 392, 71-75.

4 Davis, E., Hussey, R., Baum, T., Bakker, J., Schots, A., Rosso, M.-N. and Abad, P. (2000) Annu. Rev. Phytopathol. 38, 365-396.

5 Davis, E., Hussey, R. and Baum, T. (2004) Trends Parasitol. 20, 134-141.

6 Hussey, R., Davis, E. and Baum, T. (2002) Braz. J. Plant Physiol. 14, 183-194.

7 Jasmer, D., Goverse, A. and Smant, G. (2003) Annu. Rev. Phytopathol. 41, 245-270.

8 Vanholme, B., de Meutter, J., Tytgat, T., van Montagu, M., Coomans, A. and Gheysen, G. (2004) Gene 332, 13-27.

9 Sijmons, P., Atkinson, H. and Wyss, U. (1994) Annu. Rev. Phytopathol. 32, 235-259.

10 Gheysen, G. and Fenoll, C. (2002) Annu. Rev. Phytopathol. 40, 191-219.

11 Hussey, R. (1989) Annu. Rev. Phytopathol. 27, 123-141.

12 Wyss U. (1992) Fund. Appl. Nematol. 15, 75-89.

13 Wyss U., Grundler, F.M.W. and Muench, A. (1992) Nematologica 38, 98-111.

14 Wyss, U. and Zunke, U. (1985) Rev. Nematol. 8, 89-92.

15 Wyss, U. and Zunke, U. (1986) Rev. Nematol. 9:153-65.

16 Sundermann, C. and Hussey, R. (1988) J. Nematol. 20, 141-49.

17 Hussey, R. and Mims, C. (1991) Protoplasma 162, 99-107.

18 Cohn, E. and Mordechai, M. (1977) Phytoparasitica 5, 83-93.

19 Endo, B. (1990) in Electron Microscopy of Plant Pathogens (K. Mendgen and D.E. Lesemann, eds.), pp. 291-305. Springer Verlag, Berlin.

20 Böckenhoff, A. and Grundler, F. (1994) Parasitology 109, 249-54.

21 Urwin, P., Moller, S., Lilley, C., McPherson, M. and Atkinson, H. (1997) Mol. Plant-Microbe Interact. 10, 394-400.

22 Ray, C., Abbott, A. and Hussey, R. (1994) Mol. Biochem. Parasitol. 68, 93-101.

23 Smant, G., Stokkermans, J., Yan, Y., de Boer, J., Baum, T., Wang, X., Hussey, R., Gommers, F., Henrissat, B., Davis, E., Helder, J., Schots, A. and Bakker, J. (1998) Proc. Nat. Acad. Sci. USA 95, 4906-4911.

24 De Meutter, J., Vanholme, B., Bauw, G., Tytgat, T., Gheysen, G. and Gheysen, G. (2001) Mol. Plant Pathol. 2, 297-301.

25 Hussey, R., Paguio, O. and Seabury, F. (1990) Phytopathology 80, 709-714.

26 Vanholme, B., De Meutter, J., Tytgat, T., Gheysen, G.D.C., Vanhoutte, I. and Gheysen, G. (2001) Med. Fac. Landbouww. 66, 93-95.

27 Jaubert, S., Ledger, T.N., Laffaire, J., Piotte, C., Abad, P. and Rosso, M.-N. (2002) Mol. Biochem. Parasitol. 121, 205-211.

28 Jaubert, S., Laffaire, J., Ledger, T., Escoubas, P., Amri, E., Abad, P. and Rosso, M.-N. (2004) Int. J. Parasitol. 34, 873-880.

29 Robertson, L., Robertson, W. and Jones, J. (1999) Parasitology 119, 167-176.

30 Dautova, M., Rosso, M., Abad, P., Gommers, F., Bakker, J. and Smant, G. (2001) Nematology 3, 129-139.

31 Jones, J., Smant, G. and Blok, V. (2000) Nematology 2, 887-893.

32 Jones, J., Furlanetto, C., Bakker, E., Banks, B., Blok, V., Chen, Q., Phillips, M. and Prior, A. (2003) Mol. Plant Pathol. 4, 43-50.

33 Jones, J., Reavy, B., Smant, G. and Prior, A. (2004) Gene 324, 47-54.

34 Popeijus, H., Blok, V., Cardle, L., Bakker, E., Phillips, M., Helder, J., Smant, G. and Jones, J. (2000) Nematology 2, 567-574.

35 Popeijus, H., Overmars, H., Jones, J., Blok, V., Goverse, A., Helder, J., Schots, A., Bakker, J., and Smant, G. (2000) Nature 406, 36-37.

36 Ding, X., Shields, J., Allen, R. and Hussey, R. (1998) Mol. Plant-Microbe Interact. 11, 952-959.

37 Gao, B., Allen, R., Maier, T., Davis, E., Baum, T. and Hussey, R. (2001) Mol. Plant-Microbe Interact. 14, 1247-1254.

38 Huang, G., Dong, R., Maier, T., Allen, T., Davis, E., Baum, T. and Hussey, R. (2004) Mol. Plant Pathol. 5, 1-6.

39 Lambert, K., Allen, K. and Sussex, I. (1999) Mol. Plant-Microbe Interact. 21, 328-336.

40 Qin, L., Overmars, H., Helder, J., Popeijus, H., van der Voort, J.R., Groenink, W., van Koert, P., Schots, A., Bakker, J. and Smant, G. (2000) Mol. Plant-Microbe Interact. 13, 830-836.

41 Qin, L., Kudla, U., Roze, E., Goverse, A., Popeijus, H., Nieuwland, A., Overmars, H., Jones, J., Schots, A., Smant, G., Bakker, J. and Helder, J. (2004) Nature. 427, 30.

42 Tytgat, T., Vanholme, B., De Meutter, J., Claeys, M., Couvreur, M., Vanhoutte, I., Gheysen, G., Van Criekinge, W., Borgonie, G., Coomans, A. and Gheysen, G. (2004) Mol. Plant-Microbe Interact. 17, 846-852.

43 Gao, B., Allen, R., Maier, T., McDermott, J., Davis, E., Baum, T. and Hussey, R. (2002) Int. J. Parasitol. 32, 1293-1300.

44 Gao, B., Allen, R., Maier, T., Davis, E., Baum, T. and Hussey, R. (2003) Mol. Plant-Microbe Interact. 16, 720-726.

45 Gao, B., Allen, R., Davis, E., Baum, T. and Hussey, R. (2004) Int. J. Parasitol. 34, 1377-1383.

46 Wang, X., Allen, R., Ding, X., Goellner, M., Maier, T., de Boer, J., Baum, T. Hussey, R. and Davis, E. (2001) Mol. Plant-Microbe Interact. 14, 536-544.

47 Huang, G., Gao, B., Maier, T., Davis, E., Baum, T. and Hussey, R. (2003) Mol. Plant-Microbe Interact. 16, 376-381.

48 Huang, G., Dong, R., Allen, R., Davis, E., Baum, T. and Hussey, R. (2005) Int. J. Parasitol. 35, 685-692.

49 Huang, G., Dong, R., Allen, R., Davis, E., Baum, T. and Hussey, R. (2005) Mol. Plant Pathol. 6, 23-30.
50 Rosso, M.-N., Favery, B., Poitte, C., Arthaud, L., de Boer, J., Hussey, R., Bakker, J., Baum, T. and Abad, P. (1999) Mol. Plant-Microbe Interact. 12, 585-591.
51 de Boer, J., McDermott, J., Davis, E., Hussey, R., Popeijus, H., Smant, G. and Baum, T. (2002) J. Nematol. 34, 9-11.
52 Yan, Y., Smant, G., Stokkermans, J., Qin, L., Helder, J., Baum, T., Schots, A. and Davis, E. (1998) Gene 220, 61-70.
53 Yan, Y., Smant, G. and Davis, E. (2001) Mol. Plant-Microbe Interact. 14, 63-71.
54 Doyle, E. and Lambert, K. (2002) Mol. Plant-Microbe Interact. 15, 549-556.
55 Jaubert, S., Laffaire, J., Abad, P. and Rosso, M.-N. (2002) FEBS Lett. 522, 109-112.
56 De Boer, J., Yan, Y., Wang, X., Smant, G., Hussey, R., Davis, E. and Baum, T. (1999) Mol. Plant-Microbe Interact. 12, 663-669.
57 Gao, B., Allen, R., Davis, E., Baum, T. and Hussey, R. (2004) Mol. Plant Pathol. 5, 93-104.
58 Goellner, M., Smant, G., De Boer, J., Baum, T. and Davis, E. (2000) J. Nematol. 32, 154-165.
59 Chen, Q., Rehman, S., Smant, G. and Jones, J. (2005) Mol. Plant-Micobe Interact. 18, 621-625.
60 Shpigel E., Roiz, L., Goren, R. and Shoseyov O. (1998) Plant Physiol. 117, 1185-1194.
61 Uehara, T., Kushida, A. and Momota Y. (2001) Nematology 3, 335-341.
62 Kikuchi, T., Shibuya, H. and Jones, J. (2005) Biochem. J. 389, 117-125.
63 Doyle, E. and Lambert, K. (2003) Mol. Plant-Microbe Interact. 16, 123-131.
64 Bekal, S., Niblack, T. and Lambert, K. (2003) Mol. Plant-Microbe Interact. 16, 439-446.
65 Lambert, K., Bekal, S., Domier, L., Niblack, T., Noel, G. and Smyth, C. (2005) Mol. Plant-Microbe Interact. 18, 593-601.
66 Callis, J., Raasch, J. and Vierstra, R. (1990) J. Biol. Chem. 265, 12486-12493.
67 Lindsey, K., Casson, S. and Chilley, P. (2002) Trends Plant Sci. 7, 78-83.
68 Fletcher, J., Brand, U., Running, M., Simon, R. and Meyerowitz, E. (1999) Science 283, 1911-1914.
69 Olsen, A. and Skriver, K. (2003) Trends Plant Sci. 8, 55-57.
70 Wang, X., Mitchum, M., Gao, B., Li, C., Diab, H., Baum, T., Hussey, R. and Davis, E. (2005) Mol. Plant Pathol. 6, 187-191.
71 Goverse, A., van der Voort, J., Rouppe van der Voort, C., Kavelaars, A., Smant, G., Schots, A., Bakker, J. and Helder, J., (1999) Mol. Plant-Microbe Interact. 12, 872-881.
72 Qin, L. (2001) Molecular Genetic Analysis of the Pathogenicity of the Potato Cyst Nematode *Globodera rostochiensis*, Thesis Wageningen, The Netherlands.

73 Ding, X., Shields, J., Allen, R. and Hussey, R. (2000) Int. J. Parasitol. 30, 77-81.
74 Fang, K., Vitale, M., Fehlner, P. and King, T. (1988) Proc. Nat. Acad. Sci. U.S.A. 85, 895-899.
75 Zhan, B., Liu, Y., Badamchian, M., Williamson, A., Feng, J., Loukas, A., Hawdon, J.M. and Hotez, P.J. (2003) Int. J. Parasitol. 33, 897-907.
76 Nakhasi, H., Pogue, G., Duncan, R., Joshi, M., Atreya, C., Lee, N. and Dwyer, D. (1998) Parasitol. Today 14, 157-160.
77 Pritchard, D., Brown, A., Kasper, G., Mcelroy, P., Loukas, A., Hewitt, C., Berry, C., Füllkrug, R. and Beck, E. (1999) Parasite Immunol. 21, 439-450.
78 Fioretti, L., Warry, A., Porter, A., Haydock, P. and Curtis, R. (2001) Nematology 3, 45-54.
79 Bird, A. and Bird, J. (1991) The structure of nematodes, Academic Press, San Diego, CA.
80 Waetzig, G., Sobczak, M. and Grundler, F. (1999) Nematology 1, 681-686.
81 Robertson, L., Robertson, W., Sobczak, M., Helder, J., Tetaud, E., Ariyanayagam, M., Ferguson, M., Fairlamp, A. and Jones, J. (2000) Mol. Biochem. Parasitol. 111, 41-49.
82 Prior, A., Jones, J., Blok, V., Beauchamp, J., McDermott, L., Cooper, A. and Kennedy, M. (2001) Biochem. J. 356, 387-394.
83 Semblat, J., Rosso, M., Hussey, R., Abad, P. and Castagnone-Sereno, P. (2001) Mol. Plant-Microbe Interact. 14, 72-79.
84 Wang, X., Meyers, D., Yan, Y., Baum, T., Smant, G., Hussey, R. and Davis, E. (1999) Mol. Plant-Microbe Interact. 12, 64-67.
85 De Boer, J., Yan, Y., Smant, G., Davis, E., and Baum, T. (1998) J. Nematol. 30, 309-312.
86 Goverse, A., Davis, E., and Hussey, R. (1994) J. Nematol. 26, 251-259.
87 De Boer, J., McDermott, J., Wang, X., Maier, T., Qui, F., Hussey, R., Davis, E. and Baum, T. (2002) Mol. Plant Pathol. 3, 261-270.
88 Timmons, L and Fire, A. (1998) Nature 395, 854.
89 Rosso, M.-N., Dubrana, M., Cimbolini, N., Jaubert, S. and Abad, P. (2005) Mol. Plant-Microbe Interact. 18, 615-620.
90 Fanelli, E., Di Vito, M., Jones, J. and De Giorgi, C. (2005) Gene 349, 87-95.
91 Urwin, P., Lilley, C. and Atkinson, H. (2002) Mol. Plant-Microbe Interact. 15, 747-752.
92 Wesley, S., Helliwell, C., Smith, N., Wang, M., Rouse, D., Liu, Q., Gooding, P., Singh, S., Abbott, D., Stoutjesdijk, P., Robinson, S., Gleave, A., Green, A. and Waterhouse, P. (2001) Plant J. 27, 581-590.
93 De Meutter, J., Tytgat, T., Witters, E., Gheysen, G., Van Onckelen, H. and Gheysen, G. (2003) Mol. Plant Pathol. 4, 271-277.
94 Weerasinghe, R., Mck. Bird, D. and Allen, N. (2005) Proc. Nat. Acad. Sci. U.S.A. 102, 3147-3152.

MUTAGENESIS OF HUMAN p53 TUMOR SUPPRESSOR GENE SEQUENCES IN EMBRYONIC FIBROBLASTS OF GENETICALLY-ENGINEERED MICE

Zhipei Liu, Djeda Belharazem, Karl Rudolf Muehlbauer,
Tatiana Nedelko, Yuri Knyazev and Monica Hollstein

Department of Genetic Alterations in Carcinogenesis
German Cancer Research Center
Im Neuenheimer Feld 280
D-69120 Heidelberg
Germany

INTRODUCTION

Mutagenesis assays in mammalian cells and prokaryotic organisms were indispensable tools contributing to the elucidation of basic principles and molecular events underlying sequence changes in DNA. Two facts emanating from decades of research on molecular mechanisms of mutagenesis, e.g., with the *E. coli lacI* system, or by characterizing HPRT (hypoxanthine phosphoribosyltransferase) mutations in rodent cells and human fibroblasts, is that the chemical properties of the DNA bases and the base sequence context, in addition to biological selection, are crucial determinants of both spontaneous and carcinogen-induced mutation spectra (1-3). A mutation assay that would allow induction and selection of tumorigenic point mutations in human p53 tumor suppressor gene sequences, a major target of mutations in development of human cancer (4, 5), would be valuable because with such an experimental system it would be possible to test various hypotheses on the origins of disease-causing sequence changes, and to compare directly the human tumor p53 mutation spectra in the IARC

Genetic Engineering, Volume 28, Edited by J. K. Setlow

database (6, 7) with experimentally-generated mutation patterns . The keystone in design of a mutation test is the strategy that permits recovery of the cell, the organism, or the plasmid that harbors a mutation in a chosen target sequence. Selection typically is accomplished by manipulating growth conditions or other parameters such that all entities that are not mutated are lost, that is, they do not survive or are not retrieved. In the widely used HPRT locus mutation protocol, the rare cell with a dysfunctional mutation in the HPRT gene is recovered from the population of wild-type cells by cultivating the cells in a selection medium containing drugs that are toxic only to those cells (the vast majority) still harboring a functional HPRT locus. Development of a comparable mammalian mutation assay with human p53 sequences as the mutagen target offers special challenges, such as the task of devising a successful strategy to separate cells in culture that have p53 mutations away from cells that do not, and foster their proliferation.

To achieve this goal, we took advantage of the propensity of mouse, but not human, fibroblasts to undergo immortalization when the p53 locus is functionally inactivated (discussed below), and employed gene-targeting technology to introduce the exact human p53 sequence into mouse fibroblasts. In this way, it is possible to have a human sequence as mutagen target in mouse cells. Thus, we have combined an advantage of working with mouse cells (ability to immortalize by p53 point mutation) with the advantage of having a human p53 target sequence in the mutation assay (i.e., the exact sequence in human cells most frequently mutated in human cancers). Mouse and human p53 are highly similar, especially at the level of amino acid sequence, but due to the third wobble base of the genetic code, divergence of the DNA sequences between the 2 species is about 15%.

First, we created a human p53 knock-in mouse strain by constructing a gene-targeting vector that has the human p53 DNA-binding domain-encoding sequence. The plasmid also harbors a *loxP*-flanked neomycin phosphotransferase gene for selection of homologous recombinant mouse embryonic stem cells that can be removed subsequently by cre recombinase *in vivo* or *in vitro*. (Figure 1, and ref. 8). In the **hu**man **p**53 **k**nock-**i**n (Hupki) mice derived from the ES cells in which proper targeting by the vector occurs, exons 4-9 of endogenous mouse p53 allele are replaced with the corresponding human p53 gene sequences. Exons 4-9 encompass the segment in which most human tumor mutations arise. We included the adjacent polyproline domain (PPD) in the exchanged DNA segment also, so that the Hupki model could serve additionally to investigate the role of PPD polymorphic variants in the human population on p53 function *in vivo* (discussed in ref. 9).

The homozygous Hupki strain is phenotypically "wild-type": the mice develop normally, and do not show the biochemical and biological abnormalities that have been reported in p53 deficient and p53 knockout mice. The chimaeric transcript of the Hupki p53 gene is correctly spliced, present at normal levels in various (murine) tissues, and encodes protein that binds to p53 consensus sequences. The Hupki p53 protein also has the biological properties of functional, normal p53, such as the ability to accumulate following stress, and to activate transcriptionally known p53 target genes controlling cell-cycle checkpoints and apoptosis (8, 10, 11). The kinetics and dose response of gamma

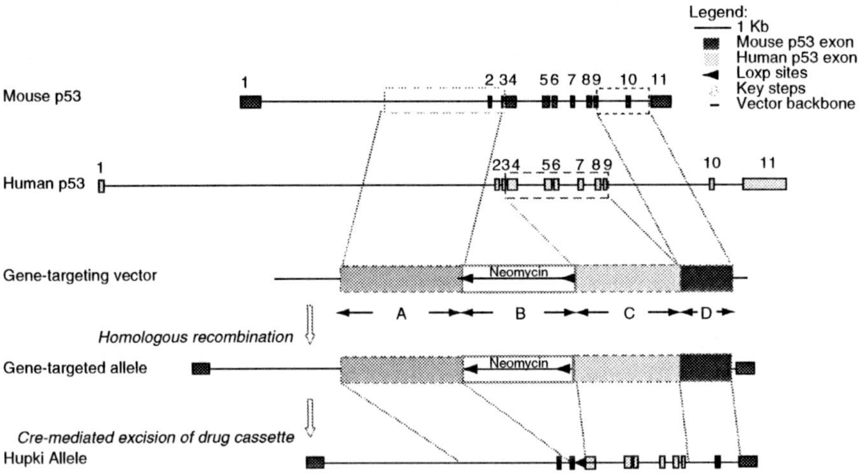

Figure 1. Gene-targeting strategy generating Hupki strains.

irradiation-triggered apoptosis in thymocytes, which is a strictly p53-dependent process (12, 13), are similar in Hupki mice and mice with a normal mouse p53 gene (8). Since typical p53 wild-type responses to DNA damage and other apoptotic stimuli are intact in Hupki prototype mice, the strain and its mutant derivatives can be applied to the study of human p53 PPD and DBD structural/functional properties *in vivo*. Second, the Hupki mouse model provides a unique experimental tool for elucidation of human tumor p53 mutation spectra, both *in vivo* and in Hupki fibroblasts *in vitro*. This latter application and its ramifications are the subject of this chapter.

IMMORTALIZATION MECHANISMS OF MURINE EMBRYONIC FIBROBLASTS(MEFs): CRUCIAL ROLE OF p53 POINT MUTATIONS

Under standard cell-culture conditions, primary murine embryonic fibroblasts from wild-type mice stop proliferating after 10 or more population doubling. At this stage, cells become senescent, acquiring an irregular, and often enlarged, flattened morphology. Functional inactivation of the ARF/p53 tumor suppressor pathway allows cells to bypass proliferation block (14-16). Cell division is resumed, and the recovering cell population regains a homogeneous morphology. Spontaneous bypass of senescence and death is a relatively rare event ($<1/10^6$) in murine embryonic fibroblast (MEF) cultures; nevertheless, spontaneous immortalization of primary mouse cells is orders of magnitude more common than immortalization of human cells (15, 17). Inactivation of the p19ARF-p53 pathway, but not p16INK4a, appears to be sufficient for MEF immortalization, which typically occurs by mutation of p53 or by loss of INK4a/ARF gene sequences (14, 18-20). The INK4a/ARF locus encodes two distinct tumor suppressors, p16INK4a and p19ARF. These two genes have a different transcription start but share exons 2 and 3, processed in different reading frames, thus encoding proteins with unrelated amino acid sequences (21). p16INK4a is a core component of the cell-cycle control machinery and responds to both positive and negative growth regulatory signals. It regulates the activities of CyclinD-Cdk4/6 complex, and consequently affects pRB tumor suppressor function and E2F responsive genes (22). ARF, the alternative reading frame product, is a key mediator of p53-dependent growth suppression. It exists at a low or undetectable level in most normal cell and tissues types (23). ARF can be directly or indirectly induced in response to abnormal proliferation signals, such as continued *in vitro* culturing (19), and inappropriate expression of proliferative oncogenes, including Ras (24, 25), c-myc (26) and E2F1 (27). Although ARF is thought to function mainly as an activator of p53 by neutralizing the activities of Mdm2, ARF has other p53-independent functions impinging on growth control and apoptotic decisions of cells (28, 29).

HUF (HUPKI EMBRYONIC FIBROBLAST) IMMORTALIZATION AS A METHOD TO SELECT FOR DYSFUNCTIONAL POINT MUTATIONS IN HUMAN P53 GENE SEQUENCES

HUFs (Hupki embryonic fibroblasts), like MEFs, are mouse cells, and immortalize readily as anticipated. Under normal culture conditions, the serially passaged embryonic fibroblast cultures become senescent by passage 5, but then almost all cultures (<90%) recover, generating immortalized cell lines when the protocol described here is followed (Figure 2 Flow Diagram and Short Protocol below, page opposite). At passage 3, and again after the senescent phase (usually >passage 8), the morphology becomes uniform and cell doubling time is short (1-2 d) (Figure 3, Panels A and C), at passage 5-6 , however, as senescent features develop in the population, division stops for up to several weeks, cells are sparse and become flattened or irregular in shape (Figure 3 Panel B). When the primary cultures are not exposed to mutagen at early passage, but instead are allowed to immortalize spontaneously according to the protocol below, approximately 10% of the cell lines recovered will harbor a p53 mutation (11, 30, 31 and unpublished observations). When we expose primary cells at passage 1-2 to a carcinogenic mutagen (UV light, aristolochic acid, benzo(a) pyrene), up to 40% of the recovered immortalized cell lines can harbor one or even 2 p53 mutations. The mutagen-induced *p53* mutations in HUFs we have characterized thus far display various features of human tumor *p53* mutations recorded in the IARC database (6, 7). For example, cell lines recovered from cultures initially exposed to the tobacco carcinogen benzo(a)pyrene frequently harbor a p53 gene G to T transversion on the non-transcribed strand (11), as do lung tumors of smokers (6). To optimize screening of immortalized HUF cell lines for the presence of mutant p53, high throughput procedures can be performed, such as the p53GeneChip protocol developed by Affymetrix (30, 32).

The protocol we currently use for performing mutagenesis experiments with HUFs is as follows (see also Figure 2, and accompanying footnotes).

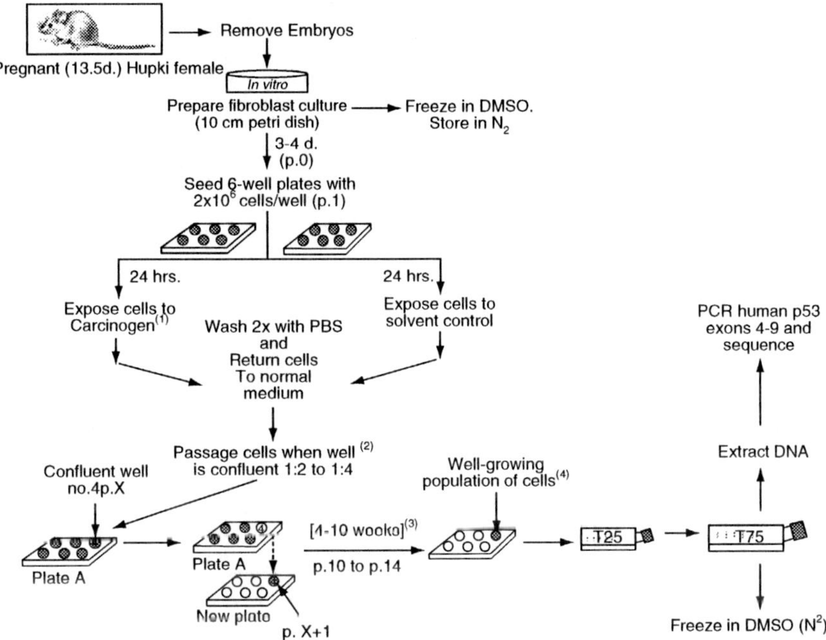

Figure 2. P53 mutation assay in HUFs: Flow Diagram. Notes: (1) Various treatment times are possible, e.g., from hours to days and can be repeated during early passages. (2) When the well is fully confluent, even at the periphery of the well, we split 1:4. When the confluent area is concentrated in the center of the dish, we split 1:2. (Early passage cells tend to migrate towards the center of the well over time.) The passaging conditions are critical because prior to immortalization, sparse cultures may not survive. (3) At passage 5-6, cells become senescent and the cultures can remain subconfluent for several weeks.

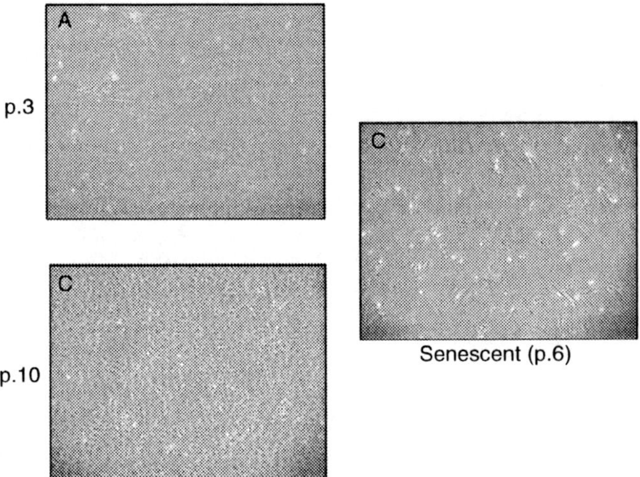

Figure 3. Photomicrographs of HUF cells (Courtesy of J. vom Brocke). A. Primary cells at passage 3. B. A senescent culture at passage 6. C. An immortalized cell culture, passage 10.

P53 MUTAGENESIS ASSAY: SHORT PROTOCOL

Note: For guidance on isolation of primary embryonic fibroblasts refer to the procedure for preparation of MEF feeder layers, Chapter 13 of Torres & Kuhn 1997 (33):

1. Sacrifice pregnant females at day 13.5 of pregnancy. Using sterile technique, dissect out the embryos, place in sterile PBS, and remove head, spleen and liver of each embryo.
2. Mince the embryo with fine scissors.
3. Add 1-2 ml of sterile medium, and pass tissue through a 20G needle 10X, then through a 25G needle 5X.
4. Transfer cell suspensions to a 10-cm dish containing 10 ml medium (cells from 1-2 embryos per dish). Medium: DMEM with 10% FCS, supplemented with penicillin and streptomycin, L-glutamine, and sodium pyruvate. See Torres & Kuhn 1997 (33).
5. Incubate the cells in a tissue culture incubator (37 °C, 5% CO_2), changing the medium daily to discard floating cells and debris. When cells are confluent, trypsinize and transfer cells to two 15-cm dishes containing 25 ml fresh medium.
6. When confluent: a) freeze cells in DMSO and store in liquid nitrogen for future experiments, and/or b) proceed with setup of mutagenesis experiment.
7. Seed 6-well plates with (1 to) 2 X 10^5 cells per well*. Label each well and plate. Prepare plates destined for treatment, and for solvent control. (*For >24 hr treatment protocols, we seed a lower number of cells).
8. On the following day, remove medium and replace with carcinogen- or solvent-containing medium. Incubate the cells for the treatment time chosen.
9. Remove medium, wash cells 2 X with sterile PBS, then pipet fresh medium into all the wells.
10. On the following day, or when a well becomes confluent, trypsinize and passage cells at 1:2 or 1:4 depending on the degree of confluency. Evaluate and handle each well separately.
11. By passage 5 the cells usually have stopped growing, the monolayer becomes sparse, and the cells acquire irregular and often enlarged morphologies. In this senescent stage, which can last for several weeks, the cultures are not passaged. Medium is changed 1X per week.
12. As growth resumes, and a well acquires large areas of confluency or is confluent, resume serial passaging of the cells. Each well is an independent separate culture.
13. When a culture in a well again becomes confluent within several days following a passaging at 1:4, all the cells in the well can be transferred to a T25 culture flask.
14. When the T25 flask is confluent, transfer all the cells to a T75 flask. When this flask is confluent, freeze half the cells in DMSO medium for safe-keeping and store in liquid nitrogen. Continue passaging the remaining cells until passage number reaches at least 15. These cultures then are considered immortalized and we refer to them as immortalized HUF cell lines.
15. Extract DNA from an aliquot of cells from each immortalized cell line. Amplify p53 exons individually (4-9) using the primers in Table 1A. Purify

Table 1A. PCR primers for p53 sequencing from genomic DNA.

Amplicon	Primer	Sequence	Product size
Exon 4	GCEx4F	GTCCTCTGACTGCTCTTTTCACCCATCTAC	368 bp
	GCEx4R	GGGATACGGCCAGGCATTGAAGTCTC	
Exon 5	GCEx5F	CTTGTGCCCTGACTTTCAACTCTGTCTC	272 bp
	GCEx5R	TGGGCAACCAGCCCTGTCGTCTCTCCA	
Exon 6	GCEx6F	CCAGGCCTCTGATTCCTCACTGATTGCTC	204 bp
	GCEx6R	GCCACTGACAACCACCCTTAACCCCTC	
Exon 7	GCEx7F	GCCTCATCTTGGGCCTGTGTTATCTCC	175 bp
	GCEx7R	GGCCAGTGTGCAGGGTGGCAAGTGGCTC	
Exon 8	GCEx8F	GTAGGACCTGATTTCCTTACTGCCTCTTGC	241 bp
	GCEx8R	ATAACTGCACCCTTGGTCTCCTCCACCGC	
Exon9	GCEx9F	CACTTTTATCACCTTTCCTTGCCTCTTTCC	146 bp
	GCEx9R	AACTTTCCACTTGATAAGAGGTCCCAAGAC	

the PCR products with Microcon filters and perform dideoxy sequencing of each fragment. (Alternatively, multiplex PCR can be performed, and PCR products fragmented, labeled, and analyzed with Affymetrix p53GeneChip microarrays as described by Liu et al., 2004, ref. 30).

INDICATORS OF LOSS OF p53 FUNCTION IN HUF CELL LINES

p53 generally is kept at low levels in normal cells and tissues. Under certain kinds of stress, such as DNA damage, p53 is stabilized and accumulates in the nucleus. The level and kinetics of p53 accumulation are carefully controlled by a series of positive and negative feedback loops (34-36). Basal p53 protein level can also change if the *p53* gene becomes mutated. p53 structure and function are highly sensitive to a myriad of single amino acid changes in protein sequence that destroy p53 function and disturb the control of p53 stability (37, 38). In the case of a missense mutation, the dysfunctional p53 protein typically accumulates to abnormally high levels. However, in the case of nonsense mutations and most of the splicing mutations, p53 protein is usually absent due to nonsense-mediated RNA decay or the instability of truncated p53 proteins.

Under normal (unstressed) conditions, primary HUFs (pHUFs) and HUF cell lines with wild-type unmutated (WT) *p53* display weak nuclear staining when incubated with antiserum CM1 against human p53, just as their wild-type (WT) MEF counterparts do when treated with mouse anti p53 antiserum. However, when cells are treated with UVC (30 J/m^2, 12 hrs after irradiation), or with adriamycin (1 microM for 12 hrs), and subsequently stained, an intense signal is observed in the nuclei. This is in keeping with the known wild-type p53 response to DNA damage. Immunocytochemical staining of treated or untreated pHUFs with normal rabbit IgG is used as a negative control for specificity of staining. Cellular p53 protein levels can also be detected conveniently by immunoblot analysis.

Immortalized HUF cell lines harboring missense mutant p53, but not HUF cell lines with unmutated (i.e., WT) p53, stain intensely without prior exposure to a DNA damaging agent. HUF cell lines that do not produce p53 protein due to a nonsense or frameshift mutation do not show this staining; instead, they stain similarly to cells with wild-type p53, as expected. However, they can be distinguished in their staining pattern from nonmutant (WT) cell lines by treating the cells with damaging agent first: WT cells will show strong staining because p53 is induced, whereas p53 null cells remain immunohistochemically negative. A first rapid appraisal of *p53* gene status in HUFs thus can be obtained by examining p53 protein basal level and/or lack of the typical wild-type response to DNA damage (i.e., p53 nuclear accumulation).

Another preliminary means to identify HUF cell lines with missense mutations is to extract RNA from the culture, amplify the p53 transcript from cDNA with specific p53 primers in a single PCR reaction (Table 1B), and sequence. A screen for coding region inactivating mutations in p19/ARF exons can be performed in similar fashion (ARF-specific primers in Table 1B).

PERSPECTIVES

One of the reasons for establishing the Hupki mouse strain was to provide an experimental tool for investigating mutagenic activity of carcinogens that employs the human *p53* gene DBD as target sequence. In this chapter we describe an assay utilizing Hupki embryonic mouse fibroblasts (HUFs) to select for mutations in human p53 sequences induced by carcinogens *in vitro*. Characterization of mutations that have arisen due to pro-mutagenic conditions such as oxidative stress or as a consequence of DNA repair defects may also be feasible with this assay. Frequent detection of HUF cell lines with *p53* mutations derived from primary cultures exposed to mutagenic carcinogens confirms that *p53* mutation is a key component of *in vitro* immortalization of HUFs.

Generation of mutation spectra in HUF cells could be facilitated by further development and experimentation with the assay. For example, novel sequencing methods sensitive enough to detect a small population of mutated cells among large numbers of wild-type cells could be applied a few weeks after mutagen treatment to accelerate discovery of mutations. This would reduce the waiting time to mutation screening (typically 3 months), otherwise needed to allow the number of mutant cells to expand, take over the culture, and become a cell line. A second variant would be to determine the protocol conditions that maximize chances for recovery of immortalized cells harboring defects in p53 rather than defective p19ARF (19, 26). Further studies of genetic mechanisms that lead to HUF immortalization also could lead to discovery of strategies that favor selection of cells with mutant p53. Subcloning of cell populations, in which mutant p53 cells may reside as a small subpopulation, although too labor-intensive to be

Table 1B. PCR primers for p53 and p19ARF sequencing from cDNA.

Amplicon	Primer	Sequence	Product size
p53	mp53e2F	ATGACTGCCATGGAGGAGTC	1.2 kb
Exon 2-11	mp53e11R	TCAGGCCCCACTTTCTTGAC	
p19ARF	p19-F1	CTTGGTCACTGTGAGGATTC	568 bp
Exon 1-2	p19ex2-R1	TGAGGCCGGATTTAGCTCTGCTC	

practical as a mutation assay protocol, would undoubtedly increase the number of independent mutations identified per experiment. With the current protocol at least, it is already clear that mutations recovered do correspond remarkably well to the common p53 mutations found in human tumors (11, 30, 31).

REFERENCES

1 Miller, J.H. (1982) Cell 31, 5-7.
2 Miller, J.H. (1983) Annu. Rev. Genet. 17, 215-238.
3 Yang, J.L., Chen, R.H., Maher, V.M. and McCormick, J.J. (1991) Carcinogenesis 12, 71-75.
4 Vogelstein, B., Lane, D. and Levine, A.J. (2000) Nature 408, 307-310.
5 Harris, C.C. (1996) J. Nat. Cancer Inst. 88, 1442-1455.
6 Olivier, M., Eeles, R., Hollstein, M., Khan, M.A., Harris, C.C. and Hainaut, P. (2002) Hum. Mutat. 19, 607-614.
7 Hollstein, M., Shomer, B., Greenblatt, M., Soussi, T., Hovig, E., Montesano, R. and Harris, C.C. (1996) Nucl. Acids Res. 24, 141-146.
8 Luo, J.L., Tong, W.M., Yoon, J.H., Hergenhahn, M., Koomagi, R., Yang, Q., Galendo, D., Pfeifer, G.P., Wang, Z.Q. and Hollstein, M. (2001) Cancer Res. 61, 8158-8163.
9 Hergenhahn, M., Luo, J.L. and Hollstein, M. (2004) Cell Cycle 3, 738-741.
10 Luo, J.L., Yang, Q., Tong, W.M., Hergenhahn, M., Wang, Z.Q. and Hollstein, M. (2001) Oncogene 20, 320-328.
11 Liu, Z., Muehlbauer, K.R., Schmeiser, H.H., Hergenhahn, M., Belharazem, D. and Hollstein, M. C. (2005) Cancer Res. 65, 2583-2587.
12 Lowe, S.W., Schmitt, E.M., Smith, S.W., Osborne, B.A. and Jacks, T. (1993) Nature 362, 847-849.
13 Clarke, A.R., Purdie, C.A., Harrison, D.J., Morris, R.G., Bird, C.C., Hooper, M.L. and Wyllie, A.H. (1993) Nature 362, 849-852.
14 Harvey, D.M. and Levine, A.J. (1991) Genes Dev. 5, 2375-2385.
15 Hahn, W.C. and Weinberg, R.A. (2002) Nat. Rev. Cancer 2, 331-341.
16 Carnero, A., Hudson, J.D., Price, C.M. and Beach, D.H. (2000) Nat. Cell Biol. 2, 148-155.
17 Boehm, J.S., Hession, M.T., Bulmer, S.E. and Hahn, W.C. (2005) Mol. Cell Biol. 25, 6464-6474.
18 Serrano, M., Lee, H., Chin, L., Cordon-Cardo, C., Beach, D. and DePinho, R.A. (1996) Cell 85, 27-37.
19 Kamijo, T., Zindy, F., Roussel, M.F., Quelle, D.E., Downing, J.R., Ashmun, R.A., Grosveld, G. and Sherr, C.J. (1997) Cell 91, 649-659.
20 Sharpless, N.E., Ramsey, M.R., Balasubramanian, P., Castrillon, D.H. and DePinho, R.A. (2004) Oncogene 23, 379-385.
21 Quelle, D.E., Zindy, F., Ashmun, R.A. and Sherr, C.J. (1995) Cell 83, 993-1000.
22 Sherr, C.J. and Roberts, J.M. (1999) Genes Dev. 13, 1501-1512.
23 Zindy, F., Williams, R.T., Baudino, T.A., Rehg, J.E., Skapek, S.X., Cleveland, J.L., Roussel, M.F. and Sherr, C.J. (2003) Proc. Nat. Acad. Sci. U.S.A 100, 15930-15935.

24 Palmero, I., Pantoja, C. and Serrano, M. (1998) Nature 395, 125-126.
25 Serrano, M., Lin, A.W., McCurrach, M.E., Beach, D. and Lowe, S.W. (1997) Cell 88, 593-602.
26 Zindy, F., Eischen, C.M., Randle, D.H., Kamijo, T., Cleveland, J.L., Sherr, C.J. and Roussel, M.F. (1998) Genes Dev. 12, 2424-2433.
27 Dimri, G.P., Itahana, K., Acosta, M. and Campisi, J. (2000) Mol. Cell Biol. 20, 273-285.
28 Weber, J.D., Jeffers, J.R., Rehg, J.E., Randle, D.H., Lozano, G., Roussel, M.F., Sherr, C.J. and Zambetti, G.P. (2000) Genes Dev. 14, 2358-2365.
29 Datta, A., Nag, A. and Raychaudhuri, P. (2002) Mol. Cell Biol. 22, 8398-8408.
30 Liu, Z., Hergenhahn, M., Schmeiser, H.H., Wogan, G.N., Hong, A. and Hollstein, M. (2004) Proc. Nat. Acad. Sci. U.S.A 101, 2963-2968.
31 Feldmeyer, N., Schmeiser, H.H., Muehlbauer, K.R., Belharazem, D., Knyazev, Y., Nedelko, T. and Hollstein, M. (2006) A Cell Immortalization Assay to Investigate the Mutation Signature of Aristolochic Acid in Human p53 Gene Sequences. Mutation Res., submitted.
32 Ahrendt, S.A., Halachmi, S., Chow, J.T., Wu, L., Halachmi, N., Yang, S.C., Wehage, S., Jen, J. and Sidransky, D. (1999) Proc. Nat. Acad. Sci. U.S.A. 96, 7382-7387.
33 Torres, R.M. and Kuhn, R. (1997) Laboratory Protocols for Conditional Gene Targeting, Oxford University Press, Oxford.
34 Appella, E. and Anderson, C.W. (2001) Eur. J. Biochem. 268, 2764-2772.
35 Vousden, K.H. and Lu, X. (2002) Nat. Rev. Cancer 2, 594-604.
36 Ko, L.J. and Prives, C. (1996) Genes Dev. 10, 1054-1072.
37 Levine, A.J. (1997) Cell 88, 323-331.
38 Hainaut, P. and Hollstein, M. (2000) Adv. Cancer Res. 77, 81-137.

SALICYLIC ACID-, JASMONIC ACID- AND ETHYLENE-MEDIATED REGULATION OF PLANT DEFENSE SIGNALING

Aardra Kachroo and Pradeep Kachroo

Department of Plant Pathology
University of Kentucky
Lexington, KY 40546

INTRODUCTION

Plants being sedentary are subject to invasion by many more pathogens as compared to other mobile eukaryotes. However, these tenacious organisms have developed a vast array of defense mechanisms to ward off invasions. Since constitutive activation of defenses could compromise normal growth, plants have developed complex mechanisms to exert control over these pathways. Plants respond to pathogen invasion by activating the production of antimicrobial compounds, cell-wall reinforcement *via* the synthesis of lignin and callose and by specifically inducing elaborate defense-signaling pathways. The major players regulating the signaling pathways include the plant hormones salicylic acid (SA), jasmonic acid (JA), ethylene (ET) and to a lesser extent, abscisic acid (ABA). Each hormone activates a specific pathway and these act individually, synergistically or antagonistically, depending upon the pathogen involved (see Figure 1 for interactions between the SA, JA and ET pathways). In addition to local resistance, many of these phytohormones also induce defense responses in systemic tissues. Systemic acquired resistance (SAR) is induced in distal tissues following pathogen infection and, con-

Genetic Engineering, Volume 28, Edited by J. K. Setlow

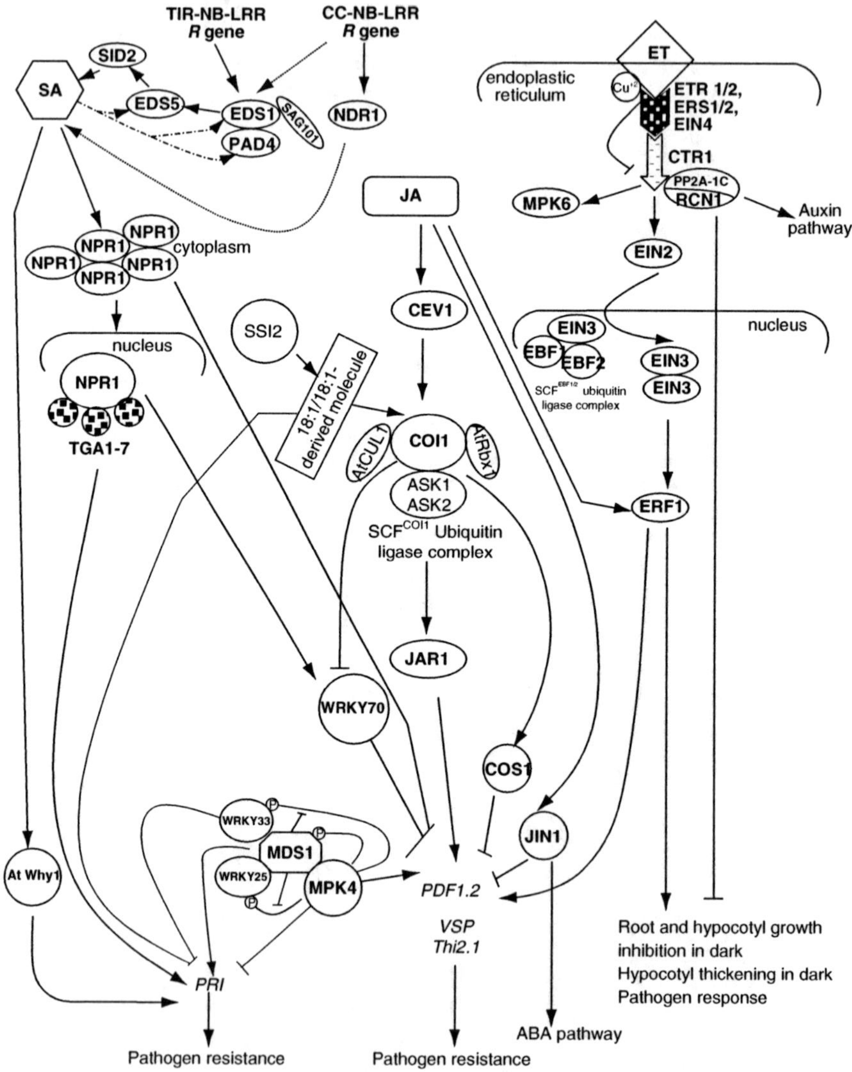

Figure 1. The figure describes the major components of the SA-, JA- and ET-mediated plant defense pathways. The defense signaling is initiated upon interaction between R and AVR proteins. Arrows indicate activation, while bars indicate repression of downstream events. Phosphorylation is indicated by an encircled P and copper ion is indicated as an encircled Cu^{+2}. Components known to interact physically are shown in contact with each other. Dotted line from NDR1 (nitrous oxide) to SA indicates partial enhancement of SA levels *via* NDR1. Dotted line from CC-NB-LRR type R gene to EDS1 indicates uncommon signaling event. Dash and dotted line indicates feedback regulation.

fers a long-lasting resistance to secondary infections by a broad spectrum of pathogens (1), while induced systemic resistance (ISR) is triggered upon root colonization by some non-pathogenic species of rhizobacteria (2). Although this chapter will focus on the involvement of SA, JA and ET in defense signaling, it is imperative to remember that these hormones are also involved in the normal physiological processes of the plant. For instance, in addition to participating in the biosynthesis of secondary metabolites (3-6), JA is also required for maintaining male fertility in the plant (4, 7). Similarly, ET is involved in many developmental processes including senescence, abscission, fertilization, fruit ripening, and seed germination (8-10). Indeed, the evidence for interconnections between normal developmental and defense-signaling pathways in plants is mounting.

EARLY EVENTS FOLLOWING PATHOGEN PERCEPTION

Perception of pathogen invasion usually involves the recognition of a pathogen-specific molecule (an avirulence or AVR factor) by a corresponding receptor molecule (a resistance or R protein) in the plant. The nature of the interaction, as well as the type of R protein, determines the particular defense pathway(s) to be activated. Recognition is often followed by the rapid generation of reactive oxygen species (ROS), which are often associated with a hypersensitive reaction (HR) resulting in cell death at the localized region of infection. HR is thought to limit the spread of the pathogen. Recognition of a pathogen is also followed by the activation of downstream defense-related events, the strength and timing of which determine the outcome of the host-pathogen interaction. Incompatible interactions (resistance response, when the pathogen AVR protein is recognized by the plant R protein) mount a more rapid and stronger response as compared to compatible interactions (susceptible response, when the plant R protein is unable to recognize the corresponding avr protein).

The oxidative burst leading to the generation of superoxide (O_2^-) and, subsequently, H_2O_2, is one of the earliest events preceding HR. These ROS are antimicrobial and induce oxidative cross-linking of the cell wall, providing a barrier for pathogen entry (11). They are also responsible for the upregulation of defense-related genes (12). Experiments linking the generation of H_2O_2 to SAR have shown that ROS may be involved in integrating a wide array of local and systemic-defense responses (13). The SA-induced activation of the oxidative burst suggests possible cross-talk between the ROS and SA-mediated pathways (14-17). Furthermore, SA and nitric oxide (NO) inhibit the activity of ROS-detoxifying enzymes, such as ascorbate peroxidase and catalase (18). This suppression is important for the onset of programmed cell death involved in HR (19, 20). ROS-mediated defense gene induction possibly affects the JA pathway as well. This is evident in the *ocp3* (overexpression of cationic peroxidase) mutant of Arabidopsis, which exhibits constitutive expression of an H_2O_2 inducible cationic peroxidase, *Ep5C* (21). The *ocp3* mutant produces increased amounts of H_2O_2, constitutively expresses the plant defensin1.2 (*PDF1.2*) gene, and exhibits enhanced resistance to the necrotrophic pathogens *Botrytis cinerea* and *Plectospaerella cucumerina*. Epistasis analysis with defense-related mutants and the observation that the *OCP3* gene encodes a transcription factor suggest that OCP3 may be involved in the regulation of the JA pathway (22).

Although many factors participate in the regulation of the complex and interconnected defense pathways in plants, this chapter will focus on the SA-, JA- and ET-mediated regulation of defense against microbes. Table 1 provides details of the various mutants/genes involved in these different pathways that have been described in this chapter.

Table 1. Arabidopsis genes involved in SA-, JA-, FA and ET-mediated defense signaling pathways.

Mutant/Gene	Locus	Protein features	Role in signaling
Salicylic Acid (SA) pathway			
SID2	AT1G74710	(ICS1) Isochorishmate synthase	Involved in SA biosynthesis
EDS5	AT4G39030	Member of the MATE transporter family, SA biosynthesis	Contributes to SA production Mutants defective in SAR and hypersusceptible to biotrophic pathogens
EDS1	AT3G48090	Lipase-like	Positive regulator of SA pathway. Binds to PAD4 and SAG101. Mutation suppresses *R* gene-mediated resistance to *Peronospora parasitica*.
PAD4	AT3G52430	Lipase-like	Camelexin biosynthesis gene. Acts downstream of *R* genes and transduces signals is response to change in redox state. Binds to EDS1.
NDR1	AT3G20600	Membrane-associated protein	Required for non-race-specific resistance to bacterial and fungal pathogens. Involved in SAR.
SAG101		Acyl hydrolase	Protein binds to EDS1. Involved in senescence and basal resistance to oomycete pathogens.
NPR1	AT1G64280	Ankyrin repeat containing protein	Positive regulator of SA pathway. Essential for the transduction of the SA signal.
NIMIN1	AT1G02450	Nuclear protein	NPR1 interacting protein. Negatively regulates *PR-1* gene expression.
TGA2, 5, 6	TGA2-AT5G06950 TGA5-AT5G06960 TGA6-AT3G12250	Basic leucine zipper family of transcription factors	NPR1 interacting proteins. Triple mutant is defective in SAR and *PR* induction.
AtWHY1	AT1G14410	Transcription factor	Participates in SA-dependent plant disease responses.
MKS1		Plant-specific protein involved in phosphorylation	Binds to MPK4 protein and inhibits phosphorylation of WRKY factors by MPK4. Overexpression confers increased resistance to bacterial pathogens.
SNI1	AT4G18470	Leucine-rich nuclear protein	Negatively regulates SAR and represses *PR* gene induction. Suppressor of *npr1*.

LSD1	AT4G20380	Zn-finger protein similar to GATA-type transcription factors	Monitors a superoxide-dependent signal and negatively regulates basal defense and cell death pathways.
ACD6	AT4G14400	Ankyrin repeat containing protein	Involved in cell death and resistance to *Pseudomonas syringae*.
ACD11	AT2G34690	Glycolipid transfer-like protein, has sphingosine transfer activity	Regulates cell death and SA-related defense responses.
SSI4	AT5G41750	TIR-NBS-LRR R-protein	Involved in defense response to bacterial pathogens.
EDR1	AT1G08720	Mitogen-activated-protein kinase kinase kinase, similar to CTR1	Mutation confers enhanced resistance to powdery mildew causing fungi.
Fatty acid (FA)/Lipid pathway			
FAD3, 7, 8	FAD3-AT2G29980 FAD7-AT3G11170 FAD8-AT5G05580	Omega-3-FA desaturases	Desaturation of 18:2 to 18:3, biosynthesis of the JA precursor linolenic acid. Triple mutant is defective in the accumulation of JA and resistance to soil-borne pathogens.
DIR1	AT5G48485	Lipid transfer protein	Defective in *PR* gene induction and onset of SAR in distal tissue. Possibly involved in transducing the mobile SAR signal.
**SSI2*	AT2G43710	Stearoyl-acyl carrier protein-desaturase	Involved in fatty acid metabolism, SA and JA signaling. Mutants are upregulated in SA pathway and defective in JA pathway. **Mediates cross-talk between SA and JA pathways.**
GLY1	AT2G40690	Glycerol-3-phosphate (G3P) dehydrogenase	Glycerolipid biosynthesis and SA signaling pathway. Suppresses constitutive SA signaling and restores JA signaling in *ssi2* plants.
ACT1	AT1G32200	G3P acyl transferase	Phosphatidic acid biosynthesis. Suppresses constitutive SA signaling and restores JA signaling in *ssi2* plants.
Jasmonic Acid (JA) pathway			
**MPK4*	AT4G01370	Mitogen-activated-protein kinase	Mutants defective in the JA pathway but upregulated in the SA pathway. Resistant to SA-responsive pathogens but hypersusceptible to JA/ET-responsive fungi. **Mediates cross-talk between SA and JA pathways.**
CEV1	AT5G05170	Cellulose synthetase	Negatively regulates JA signaling. Mutation results in constitutive JA and ET production.
COI1	AT2G39940	E3 ubiquitin ligase SCF complex F-box subunit	Key component of JA pathway. Involved in ubiquitination of regulators of JA pathway.

(*Continued*)

Table 1. Arabidopsis genes involved in SA-, JA-, FA and ET-mediated defense signaling pathways.—Cont'd

Mutant/Gene	Locus	Protein features	Role in signaling
COS1	AT2G44050	Lumazine synthase	Component of the riboflavin pathway and negative regulator of JA pathway. Mutation suppresses *coi1-1* related phenotypes.
JAR1	AT2G46370	JA-amino synthetase	Mutants are defective in response to jasmonates. Biochemically modifies JA by conjugating it to amino acids and activates it for signaling.
**JIN1*	AT1G32640	MYC-related transcription factor	Negatively regulates the JA pathway. Mutants are resistant to necrotrophic pathogens. Overexpression induces enhanced JA and ABA responses. **Involved in cross-talk between JA and ABA pathways.**
AtCUL1	AT4G02570	Cullin-component of SCF ubiquitin ligase complex	Associates with COI1 and other components to assemble the SCF^{COI1} complex. Mediates responses to JA and auxin.
ASK1	AT1G75950	E3-ubiquitin ligase SCF complex subunit	Associate with COI1 and AtCUL1 to assemble the SCF^{COI1} complex.
ASK2	AT5G42190	E3-ubiquitin ligase SCF complex subunit	Associate with COI1 and AtCUL1 to assemble the SCF^{COI1} complex.
Ethylene (ET) Pathway			
ETR1	AT1G66340	NH_4-ET binding domain and COOH-terminal histidine kinase domain	Involved in ET perception, mutation confers ET insensitivity and defects in induced systemic resistance.
ERS1	AT2G40940	NH_4-ET binding domain and COOH-terminal histidine kinase domain	Involved in ET perception.
ERS2	AT1G04310	NH_4-ET binding domain and COOH-terminal histidine kinase domain lacking essential catalytic amino acids	Involved in ET perception.
ETR2	AT3G23150	NH_4-ET binding domain and COOH-terminal histidine kinase domain lacking essential catalytic amino acids	Involved in ET perception.

EIN4	AT3G04580	NH_4-ET binding domain and COOH-terminal histidine kinase domain lacking essential catalytic amino acids	Involved in ET perception
CTR1	AT5G03730	Serine/threonine protein kinase. Similar to mitogen activated protein kinase kinase kinase (MAPKKK)	Negatively regulates ET signaling. Interacts with the ET receptors, ETR1, ERS2 and ETR2.
EIN2	AT5G03282	Member of the natural resistance-associated macrophage protein (NRAMP) metal transporter family	Positive regulator of ET signaling acting downstream of CTR1
EIN3	AT3G20770	Nuclear localized transcription factor	Positive regulator of ET signaling acting downstream of EIN2. Overexpression confers constitutive ET responses.
**ERF1*	AT3G23240	Member of the ERF (ethylene response factor) subfamily B-3 of ERF/AP2 transcription factor family (ERF1)	Required for a subset of ET-related responses, acting downstream of EIN3. Inducible by both ET and JA. **Involved in cross-talk between ET and JA pathways.**
**RCN1*	AT1G25490	Serine/threonine protein phosphatase 2A subunit	Protein interacts with CTR1. Mutation confers increased ET sensitivity. Positive regulator of abscisic acid-mediated signaling. **Cross-talk between ET and ABA signaling.**
EBF1	AT2G25490	F-box protein	Negative regulator of ET signaling. Involved in ubiquitin/proteosome-mediated degradation of EIN3.
EBF2	AT5G25350	F-box protein	Negative regulator of ET signaling. Involved in ubiquitin/proteosome-mediated degradation of EIN3.
EIL1	AT2G27050	Transcription factor	Overexpression confers constitutive ET responses. Mutant susceptible to necrotrophic pathogens.
EIR1	AT5G57090	Putative auxin efflux carrier similar to bacterial membrane transporters	ET signaling regulator acting similar to EIN2 also involved in response to auxin stimulus.

*Indicates gene involved in cross-talk between two different pathways.

THE SALICYLIC ACID PATHWAY

The small phenolic compound SA plays a vital role in the defense response against many pathogens. Infections with necrotizing pathogens induce increased levels of SA (23, 24) and these, in turn, induce the expression of many defense-related genes. Plants that are deficient in SA synthesis or accumulation exhibit enhanced susceptibility to pathogen infection (25-28). In Arabidopsis and tobacco, SA is also crucial for the establishment of SAR, a broad-spectrum, long-lasting resistance in the entire plant (29). SAR is associated with the generation of an unknown signal at the point of infection and transduction of this signal to distal tissues is thought to confer enhanced immunity against secondary infections. SAR is also accompanied by the induction of a set of pathogenesis-related (*PR*) genes in the systemic tissue (30). Transgenic plants overexpressing the bacterial salicylate hydroxylase (*nahG*) gene are abolished in SAR, because these plants degrade SA to catechol and are unable to accumulate SA. Furthermore, exogenous application of SA or its biologically active analogs, 2,6-dichloroisonicotinic acid (INA) and benzo(1,2,3)thiadiazole-7-carbothioic acid *S*-methyl ester (BTH), are sufficient to induce SAR in Arabidopsis and confer enhanced resistance to a variety of pathogens (25, 30).

SA can be synthesized either from phenylalanine, *via* the action of phenylalanine ammonia lyase (PAL, 31, 32), or from chorishmate *via* the shikimate pathway (33). The Arabidopsis *SID2* (SA-induction deficient) gene encodes isochorishmate synthase (ICS) and a mutation in *sid2* renders plants defective in SA synthesis and the activation of SAR (33, 34). Consequently, *sid2* plants exhibit enhanced susceptibility to oomycete and bacterial pathogens. The *SID2* gene was systemically induced upon infection of Arabidopsis with an avirulent strain of *Pseudomonas* and this induction correlated with the accumulation of SA as well as *PR* transcripts in these tissues.

Components of the SA Signaling Pathway

Besides *SID2*, the *EDS5* (enhanced disease susceptibility), *EDS1* and *PAD4* (phytoalexin deficient) encoded proteins also contribute to SA production. A mutation in *eds5* (previously called *sid1*) compromises pathogen-induced accumulation of SA and the induction of *PR-1* gene expression. Consequently this mutant is defective in the activation of SAR and hypersusceptible to infections by *Peronospora parasitica* and *Pseudomonas syringae* (35). Sequence analysis has shown that *EDS5* encodes a member of the MATE (multidrug and toxin extrusion) transporter family (36). Pathogen infection induces the accumulation of the *EDS5* transcript, in an EDS1, PAD4 and NDR1 (non-race-specific disease resistance) dependent manner. Exogenous application of SA also induces the *EDS5, EDS1* and *PAD4* transcripts, suggesting a positive feedback-regulation in the SA pathway that facilitates amplification of the defense response (37-40).

The *EDS1* and *PAD4* encoded proteins show sequence similarities to eukaryotic lipases (37, 38). A mutation in *eds1* abolishes HR and *R* gene-mediated resistance to *P. parasitica* and results in increased susceptibility to this oomycete pathogen (41). In addition, EDS1 is also required for basal resistance to virulent isolates of *P. parasitica*, *Erysiphae* and *P. syringae* (42, 43) and the onset of HR. Mutations in *pad4* result in increased susceptibility to *P. syringae* (44), reduced accumulation of SA, and of the phytoalexin, camelexin (45). The *EDS1* and *PAD4* genes generally participate in defense signaling pathways triggered by the TIR-NB-LRR (Toll-Interleukin-1 receptor/nuclear binding/leucine rich repeat) class of *R* genes (42). However, EDS1 (and PAD4) also participate in signaling triggered by *R* gene carrying a coiled coil (CC) domain at the NH_4-terminal end (40, 43). Genetic analyses suggest that EDS1 and PAD4 act as signal transducers in response to redox stress and function downstream of *R* genes (46-48). The EDS1 protein was shown to dimerize and interact with PAD4 both *in vitro* and *in vivo*, and this interaction may be important for the amplification of defense responses (49). Although *EDS1* is necessary for the pathogen-induced accumulation of *PAD4*, *EDS1* accumulation is only partially affected by a mutation in *pad4* or reduced amounts of SA (49). In addition to PAD4, EDS1 also interacts with another lipase-like protein, SAG101 (senescence-associated gene 101, 50). Similar to PAD4, SAG101 accumulation is dependent upon the presence of a functional EDS1 protein. A mutation in *sag101* disables basal resistance to virulent isolates of *P. parasitica* and this effect is more pronounced in the *pad4* background. *pad4 sag101* double mutants are also defective in TIR-NB-LRR *R* gene-mediated resistance to avirulent pathogens.

In addition to EDS1, PAD4 and EDS5, the *NDR1* encoded small, highly basic, membrane protein (51), also feeds into the SA pathway, and is required for resistance to *P. syringae* and *P. parasitica* (52). The *ndr1-1* mutant plants are compromised in the induction of SAR, and accumulate reduced SA in response to ROS production (53). In contrast to EDS1, NDR1 is generally required by the CC-NB-LRR type of *R* genes.

SA requires the function of a downstream component NPR1 (non-expressor of *PR* genes), also called NIM1 (non-inducible immunity, 54) or SAI1 (SA insensitive, 55), to trigger the expression of *PR-1* gene. A mutation in *npr1* abolishes SA-mediated induction of *PR* genes as well as SAR, suggesting that NPR1 is a positive regulator of the SA pathway (56). The NPR1 protein contains four ankyrin repeats and a nuclear localization signal (57). In the absence of SA, NPR1 exists as an oligomer *via* intermolecular disulfide bonding between the conserved cysteines (58). In its oligomeric form NPR1 remains in the cytoplasm and is sequestered from transport into the nucleus. Upon induction of SAR, plant cells accumulate antioxidants, resulting in a change in the redox state. Under these reducing conditions, the disulfide bonds between the NPR1 molecules are broken, resulting in dissociation of the NPR1 oligomer into monomers, which are then capable of transport into the nucleus. Furthermore, this monomerization of NPR1 appears to be important for subsequent *PR* gene expression (58).

Yeast two-hybrid analysis revealed that NPR1 interacts with the AHBP-1b/TGA2, TGA3, OBF5/TGA5, TGA-6 and TGA7 transcription factors belonging to the basic leucine zipper (bZIP) protein family (59-62). Interestingly, the TGA1 and TGA4 proteins interact with NPR1 only *in planta*. Both proteins contain two unique cysteines, which are responsible for intramolecular disulfide bonding. SA induction, resulting in a change in the redox state of the cell, induces reduction of the disulfide bridges allowing the proteins to interact with NPR1 and subsequently activate gene expression (63). NPR1 enhanced the binding of TGA factors to SA-responsive elements in the *PR-1* gene promoter. This DNA-binding ability of the TGA factors was disrupted in the *npr1* mutant. The *tga2 tga5 tga6* triple mutant was unable to induce *PR* gene expression in response to SA and was defective in the onset of SAR. This triple mutant was unable to induce resistance to *P. parasitica*, in spite of pretreatment with INA (64). These results demonstrate the redundancy among the TGA factors and show that these are required for NPR1-mediated *PR* gene expression.

In addition to SA signaling and SAR, NPR1 also functions in ISR and possibly in regulating cross-talk between the SA and JA pathways. It is well established that the SA and JA pathways interact antagonistically (65, 66). Activation of SAR results in the suppression of JA signaling, and SA has been shown to be an inhibitor of JA-inducible genes (67-69). The *nahG* as well as *npr1-1* plants accumulated increased amounts of JA and JA-inducible genes upon infection with *P. syringae*, as compared to wild-type plants. Exogenous application of SA inhibited the MeJA (methyl JA)-induced expression of JA-responsive genes in wild-type plants but not in *npr1-1* plants. These studies led to the suggestion that NPR1 is involved in the negative cross-talk between the SA and JA pathways (68), and the nuclear localization of NPR1 was not required for mediating this cross-talk. Thus, a novel cytosolic function of NPR1 may serve in the negative interaction between the SA and JA pathways. The onset of ISR in Arabidopsis also requires a functional NPR1 protein (70, 71), suggesting that NPR1 may be an essential signaling component of multiple signaling pathways.

The *npr1* mutant phenotypes are completely suppressed in the *sni1* (suppressor of *npr1-1* inducible) background. The *SNI1* locus encodes a leucine-rich, nuclear-localized protein. The *sni1 npr1* plants are restored in their ability to induce SAR and *PR* gene expression, as well as induced resistance to *P. syringae* and *P. parasitica*. Consequently, it has been suggested that SNI1 acts as a negative regulator of SAR by repressing *PR* gene induction in the absence of SA (72). The NIMIN1 (NIM1/NPR1-interacting) protein also appears to be a negative regulator of NPR1 functions (73). The NIMIN1 protein forms a ternary complex with NPR1 and TGA factors to bind SA-responsive promoter elements. Overexpression of the protein resulted in repression of the SA-induced expression of *PR* genes and an impairment in SAR and *R* gene-mediated resistance. In contrast, downregulation or loss of the *NIMIN1* transcript resulted in enhanced induction of *PR* genes in response to SA.

Other Transcription Factors Regulating the SA Pathway

Besides the NPR1-interacting transcription factors, other proteins including WRKY70 and AtWhy1 have also been shown to participate in the SA signal transduction pathway. Expression of the *WRKY70* gene was induced in SA and pathogen-elicitor-treated plants (74). Since basal levels of the *WRKY70* transcript were abolished in *nahG* plants, SA appears to regulate the expression of this gene. On the other hand, overexpression of *WRKY70* resulted in constitutive SA signaling. Thus, like many other regulators of the SA pathway, *WRKY70* appears to be in a feedback regulatory loop with SA (74). Antisense-suppression of *WRKY70* results in activation of COI1-dependent genes, and the transcript levels of this gene are upregulated in the *coi1* mutant. Therefore WRKY70 possibly participates in cross-talk between the SA and JA pathways.

SA induces the DNA-binding ability of AtWhy1 and this induction is absent in the *atwhy1.2* mutant. In the *npr1-1* background, SA continues to activate the DNA-binding ability of AtWhy1, suggesting that AtWhy1 functions in *PR-1* induction in an NPR1-independent manner (75). The *atwhy1* mutants are severely compromised in SA-induced resistance and exhibit enhanced susceptibility to infection by virulent as well as avirulent isolates of *P. parasitica*. These studies indicate that NPR1-independent pathways exist and that these can also contribute towards the induction of *PR* gene expression and SAR.

Constitutive Activation of SA Pathway

Several mutants constitutively accumulate high levels of SA, and these generally exhibit altered morphologies such as reduced size and/or spontaneous cell death. Examples include, the *accelerated cell death* (*acd*, 47, 76), *constitutive expressor of PR genes* (*cpr*, 77), *lesion-stimulating disease* (*lsd*, 46, 78, 79), *defense no death* (*dnd*, 80), *aberrant growth and death* (*agd*, 81), and *suppressor of salicylic acid insensitivity* (*ssi*, 39, 64, 82) mutants. Although the exact roles of many of the genes in SA signaling need to be investigated, characterization of some has provided better insights in the SA signaling pathway.

The *lsd1* mutant exhibits enhanced resistance to several virulent pathogens and uncontrolled cell death, which is dependent on EDS1 and PAD4 (46). The *LSD1* gene encodes a zinc-finger protein with similarity to GATA-type transcription factors (83). The *acd6-1* mutant exhibits constitutive defense responses, increased amounts of SA and enhanced resistance to *P. syringae*. *ACD6* encodes a novel protein with ankyrin and transmembrane domains (84). Although both LSD1 and ACD6 appear to regulate cell death and disease resistance, their exact roles in the SA pathway remain unexplored. The *acd11* mutant exhibits constitutive programmed cell death and HR-related defense gene activation. These phenotypes are dependent upon SA, EDS1 and PAD4. The ACD11 protein is a homolog of the mammalian glycolipid transfer protein and was shown to have sphingosine transfer activity (47). Thus, *ACD11* may regulate cell death and SA-related defense responses by altering sphingolipid metabolism. The *ssi4* mutant constitutively expresses *PR* genes, accumulates increased amounts of SA, and exhibits enhanced resistance to bacterial and oomycete pathogens (39). Both SA and EDS1 are required for the constitutive *PR* gene expression and enhanced disease resistance phenotypes. The *SSI4* gene encodes a TIR-NBS-LRR type of R protein, and the *ssi4* mutant phenotypes may possibly be related to the constitutive activation of this R protein.

A mutation in the *ssi2/fab2* and *mpk4* genes also results in constitutive *PR* gene expression, spontaneous lesion formation, and enhanced resistance to both bacteria and oomycete pathogens. However, unlike most other mutants studied, these also impair JA-dependent responses (82, 85). The *AtMPK4* encodes a mitogen-activated-protein kinase (MAPK). In addition to exhibiting enhanced resistance to *P. parasitica* and *P. syringae*, *mpk4* plants are also hypersusceptible to an

ET/JA pathway-inducing fungus (86), suggesting that AtMPK4 may be involved in cross-talk between the SA and JA pathways. The MPK4 protein phosphorylates a downstream substrate MKS1 (MAPK4 substrate), belonging to a class of plant-specific proteins some of which have been shown to be involved in transcriptional regulation and pathogen response (87). Overexpression of MKS1 results in increased resistance to *P. syringae*. In addition to MKS1, MAPK4 also phosphorylates the WRKY25 and WRKY33 transcription factors, but does not interact directly with these. Interestingly, MKS1 has been shown to interact physically with WRKY25 and 33, and inhibits the phosphorylation of these by MPK4, suggesting that MKS1 may regulate signaling by modulating the phosphorylation of WRKY transcription factors.

Role of Fatty Acids and Lipids in SA Signaling

Unlike MPK4, the SSI2 encoded stearoyl-ACP-desaturase (S-ACP-DES), participates in the normal metabolism of the plant and preferentially desaturates fatty acid (FA) 18:0 between carbons 9 and 10 to yield 18:1. Although S-ACP-DES catalyzes the initial desaturation step required for biosynthesis of the JA precursor, linolenic acid (18:3), a mutation in *ssi2* does not alter the levels of 18:3, perception of JA or ET, or the induced endogenous levels of JA (82). Analysis of suppressor mutants has shown that the altered phenotypes of the *ssi2* mutant are related to its reduced 18:1 levels, as opposed to the increased levels of 18:0 FA (88). A loss-of-function mutation in the gene encoding the soluble chloroplastic enzyme glycerol-3-phosphate (G3P) acyltransferase (*ACT1*) was found to revert the SA- and JA-mediated phenotypes in *ssi2* (89). A mutation in *act1* disrupts the acylation of G3P with 18:1, resulting in the accumulation of high amounts of 18:1, thereby compensating for the mutant phenotypes in the *ssi2* plants. Similarly, a mutation in the *GLY1* gene encoding the G3P dehydrogenase disrupts the formation of G3P from dihydroxyacetone phosphate and results in the restoration of 18:1 levels in the *ssi2 gly1-3* plants (90). Although the *GLY1* allele, *SFD1* (suppressor of FA desaturase deficiency) has been reported to be involved in SAR (91), experiments with the *gly1-1* mutant have thus far failed to validate its role in SAR (Venugopal, S. C. and Kachroo, P., unpublished data). Nonetheless, analysis of the *ssi2 gly1-3* mutant and the result that exogenous application of glycerol can convert wild-type plants into *ssi2*-mimics *via* reduction in 18:1 levels, strongly supports the involvement of 18:1 fatty acids (FAs) in SA-mediated signaling of the plant.

Work from other laboratories also implicates FAs and lipids in plant defense signaling. The *DIR1* (defective in induced resistance1) encoded lipid-transfer protein appears to be essential for the onset of pathogen-induced SAR (92). Although the *dir1* mutant plants are resistant to local infection by both virulent and avirulent strains of *P. syringae*, these plants are unable to induce *PR* gene expression and develop SAR in distal tissue in response to infection with virulent strains of *P. syringae* and *P. parasitica*. A closer examination has revealed that this mutant may be incapable of either producing or transducing the mobile signal essential for SAR.

In addition to DIR1, both EDS1 and PAD4 have esterase/lipase-like domains (37, 38), although the enzymatic activities of these proteins remain to be documented. An SA binding protein from tobacco has also been shown to have lipase activity (93). Together these studies suggest that FAs/lipids may form an important component of the SA signal transduction pathway.

SA-Binding Proteins

Although the molecular basis for the perception of SA remains unresolved, significant progress has been made towards identification of proteins that bind SA and possibly serve as its receptors. At least three different SA-binding proteins (SABPs) have been characterized thus far from tobacco plants and, these show varying levels of binding affinity for SA (18, 93-95). The cytosolic SABP1 (catalase) was the first SABP identified from tobacco, which bound SA in a specific manner (94). SA was shown to inhibit the catalase activity *in vitro* and it was hypothesized that such an inhibition could lead to an increase in H_2O_2 levels. SA, as well as H_2O_2, induces the expression of defense genes associated with the onset of SAR; consequently it was suggested that this catalase may relay the SA signal *via* elevating H_2O_2 levels (94). Moreover, SA and its analog INA were shown to inhibit the activity of ascorbate peroxidase, the other key H_2O_2-scavenging enzyme (96). However, other studies showed that H_2O_2 acts upstream, rather than downstream, of SA (97, 98), discounting the possibility that SABP1 (catalase) relays the SA signal *via* elevating H_2O_2 levels.

The chloroplastic SABP3 protein of tobacco was shown to have the enzymatic activity of a carbonic anhydrase, with SA-binding properties. Silencing of the carbonic anhydrase-encoding gene results in loss of the *Pto:avrPto*-mediated HR in tobacco plants (95). Among the three SABPs identified so far, SABP2 shows highest affinity for SA. Sequence analysis followed by biochemical assays, have shown that SABP2 has a strong esterase activity using methyl SA (MeSA) as substrate, and that SA is a potent inhibitor of this activity (99). RNAi (RNA interference)-mediated silencing of SABP2 resulted in enhanced local as well as systemic susceptibility to TMV (tobacco mosaic virus), suggesting that SABP2 may serve as a SA-receptor and is required for the SA-mediated defense response in tobacco (93). Furthermore, binding of SABP2 to MeSA, results in the release of SA, suggesting that SABP2 may be required to convert MeSA to SA (99). Thus, although the enzymatic functions for the SA binding proteins have been deduced, the mechanism by which they relay SA signaling still remains a mystery.

THE JASMONIC ACID PATHWAY

JA constitutes a key member of the jasmonate family of signaling molecules involved in regulating plant defense to both biotic and abiotic stresses (100, 101). Like other signaling pathways, the JA signaling pathway involves the perception of the stress stimulus, followed by local and systemic signal transduction, perception of the signal leading to the synthesis of JA and, finally, responsiveness to JA with the induction of subsequent downstream events.

One of the earliest evidence implicating JA in defense signaling came from the study of JA-biosynthesis mutants. The *fad3 fad7 fad8* triple mutant is unable to accumulate JA because it is deficient in the JA-precursor, linolenic acid (102), and is hypersusceptible to infection by insect larvae. Further examination showed that the *fad3 fad7 fad8* mutant plants are highly susceptible to root rot by *Pythium mastophorum*, and exogenous application of MeJA alleviates susceptibility to soil-born pathogens (103).

The JA signaling pathway has been well studied in the wound response of tomato plants, but this pathway appears to differ from the Arabidopsis defense pathway. For example, unlike Arabidopsis mutants, tomato plants defective in the synthesis or perception of JA are not affected in male fertility (104). Furthermore, the JA-induced systemic response pathway in tomato is shown to occur *via* the well-characterized systemin signal pathway (105). Systemin, an 18-amino acid polypeptide, and its precursor, prosystemin, act as primary signals for the activation of defense genes in systemic tissue of wounded plants (106). Antisense suppression of systemin results in increased susceptibility (107, 108), while overexpression results in increased resistance, to herbivores (109, 110). Systemin binds to a 160 kD, leucine-rich repeat-containing, receptor kinase molecule in the plasma membrane (111-113), and initiates a signal transduction cascade *via* the octadecanoid pathway (114). Early events associated with wound signaling include a rapid increase in the levels of cytosolic Ca^{+2} (115), membrane depolarization (116), inhibition of a proton ATPase in the plasma membrane (117), and the activation of MAPK activity (118). This is followed by the release of linolenic acid from membrane phospholipids by a phospholipase (119) and its subsequent conversion to 12-oxophytodienoic acid (OPDA) and JA. JA synthesized in response to wounding and systemin then induces the expression of downstream genes. Grafting experiments have shown that production of JA in the wounded tissue and perception of JA in the distal tissue is important for activation of the systemic response (120), suggesting that JA or a JA-related, octadecanoid pathway-derived molecule functions as the signal for induced systemic responses. The *spr2* (suppressor of prosystemin 2-responses) mutation impairs wound-induced JA synthesis and generation of the long-distance signal required for the induction of proteinase inhibitor genes. Map-based cloning has revealed that the *SPR2* gene encodes a chloroplastic ω3 fatty acid desaturase, involved in JA biosynthesis (121).

Components of the JA Signaling Pathway

The *COI1* gene was identified as a key component of the JA pathway and a mutation in *coi1* renders plants insensitive to growth inhibition by coronitine, a compound structurally similar to JA and methyl JA (MeJA, 4). The *coi1* plants are unable to respond to exogenous application of JA and are impaired in the induction of JA-responsive genes *VSP* (vegetative storage protein), *Thi2.1* (thionin) or *PDF1.2* (122-125). These plants are also susceptible to insect herbivory, and necrotrophic pathogens (102, 126). The *COI1* gene encodes a protein containing leucine-rich repeats and an NH_4-terminal, degenerate, F-box domain (123). Since F-box proteins in eukaryotic systems are known to function in a ubiquitin-ligase complex called the SCF (Skp1-Cdc53/Cullin-F box receptor) complex, and target substrate proteins for proteolytic degradation (127), it was predicted that COI1 may be involved in the ubiquitination of key regulators of the JA pathway. Indeed, COI1 was shown to associate with AtCUL1 (Cullin), AtRbx1 (Ring-box) and, the Skp1-like proteins, ASK1 (Arabidopsis Skp1-like) and ASK2, to assemble the SCF^{COI1} ubiquitin-ligase complex (128). Mutations in *coi1* as well as the *axr1* gene (responsible for the modification of AtCUL1) affect the SCF^{COI1} complex formation, thereby resulting in a defect in the JA response. Furthermore, the *coi1 axr1* double mutant exhibits an enhanced defect in JA response-phenotype, suggesting that the SCF^{COI1} ubiquitin-ligase complex is important for JA signaling (128). In a separate study, COI1 was shown to interact with SKP1 proteins *via* the F-box domain and with a histone deacetylase, and the small subunit of RUBISCO *via* the leucine-rich repeat domain (129). The histone deacetylase and RUBISCO could possibly serve as targets for COI1-mediated degradation and, thereby, participate in JA signaling. Based on information available from protein interactions in yeast (130), there appears to be a link between protein ubiquitination and acetylation. The histone deacetylase may be involved in the suppression of JA-responsive genes *via* association with the SCF^{COI1} complex. Furthermore, these interactions also suggest that COI1 may function in other signaling pathways. Indeed, COI1 is also required for maintaining male fertility in the plant (4).

Suppressor screens conducted using the *coi1* mutant has thus far identified one gene which when mutated suppresses all *coi1*-related phenotypes, except for male sterility (131). The *COS1* (*COI1* suppressor1) gene encodes a lumazine synthase, which is a key component of the riboflavin pathway and is required for many critical cellular processes. Since a mutation in *cos1* is able to relive the *coi1*-triggered suppression of JA-responsive genes, it has been suggested that some of the JA responses may be negatively regulated by COS1 *via* the SCF^{COI1} complex.

The *JAR1* (jasmonate resistant1) gene functions downstream of *COI1*, and is a positive regulator of JA signaling. Similar to *coi1*, a defect in *jar1* also induces defective JA responsiveness and enhanced susceptibility to necrotrophs, like *B. cinerea* (132). Protein-fold modeling of the deduced amino acid sequence of *JAR1* suggested similarities to the acyl adenylate-forming luciferase superfamily that are responsible for regulating many important cellular processes (133). Biochemical studies have demonstrated that JAR1 participates in the adenylation of JA, indicating that adenylation may be required for at least some of the JA responses (134). Further examination of the biochemical activity of JAR1 has revealed that this protein is a JA-amino synthetase and conjugates JA to several amino acids, thereby activating JA for optimal signaling (135). Since the levels of JA-isoleucine (JA-Ile) were reduced in the *jar1* mutant and because JA-Ile could inhibit root growth in the mutant, it was suggested that JA-Ile is the active form of JA.

The *jin1*, *jin4* (jasmonate-insensitive, 136) and the *jue1*, *2* and *3* (jasmonate-underexpressing, 137) mutants are also affected in JA-signaling. Although these are all JA-insensitive, they exhibit much weaker phenotypes as compared to the *coi1* mutant. This suggests that these genes may not be required for all JA-related responses. The *JIN1* gene encodes a helix-loop-helix-leucine zipper (bHLHzip-type transcription factor), designated AtMYC2 (138). The AtMYC2 protein negatively regulates the JA signaling branch required for the induction of pathogen-responsive defense genes. Interestingly and unlike the *coi1* mutant, *jin1* mutant plants do not exhibit enhanced susceptibility to necrotrophic pathogens such as *Botrytis cinerea* and *Plectosphaerella cucumerina*. Furthermore, although *jin1* plants are defective in the accumulation of some JA-induced wound-responsive genes, they accumulate increased levels of pathogen-responsive genes such as *PR-1*, *PR-4* and *PDF1.2*, upon exposure to JA (138). Consequently, overexpression of *JIN1* induces enhanced wound-related JA responses as well as ABA responses, indicating that it is involved in the cross-talk between the JA and abscisic acid (ABA) pathways. These results suggest that although JIN1 may be required for some JA-related responses, it may repress other responses such as resistance to necrotrophic pathogens.

The Arabidopsis proteins MPK4 and SSI2 are two other components required for JA-mediated signaling leading to the expression of defense genes. *MPK4* is induced within minutes after wounding (139) and it is possible that the primary stress signal or the rapidly released endogenous JA may be responsible for this activation, rather than the newly-synthesized JA. Both the *mpk4* and *ssi2* mutants are defective in the induction of JA-responsive genes, although they are unaffected in the perception of JA (83, 139). Consequently, these mutants are susceptible to necrotrophic pathogens such as *Botrytis* and *Alternaria*, respectively. The JA-mediated induction of *PDF1.2* in *ssi2 act1* plants is dependent on *COI1*, suggesting that the 18:1 signal associated with JA-responsiveness acts upstream of *COI1* (89).

The *CEV1* (constitutive expression of *VSP*) gene is another candidate, possibly participating in the perception and/or transduction of the stress signal prior to JA biosynthesis. A mutation in this cellulose synthetase-encoding gene results in the constitutive production of JA and ET and, consequently, induces the JA-responsive genes, *PDF1.2* (140), *Thi2.1* (141) and *VSP* (122). Mutant plants also exhibit enhanced resistance to powdery mildew-causing fungal pathogens (142). Because the *cev1* phenotype was only partially rescued in the *coi1* background, it appears that CEV1 acts upstream of COI1.

In addition to the *cev1* mutant, several other mutants exhibit constitutive JA responses. These include the *cex1*, *cet1/9* and *joe1/2* (137, 143, 144). The *cex1* (constitutive expression of JA-inducible genes) mutant exhibits constitutive JA-responsive phenotypes, such as JA-induced growth inhibition and the induction of JA-responsive genes (144), suggesting that CEX1 is a negative regulator of JA signaling. Genetic analysis indicates that *CEX1* acts downstream of *COI1*. The *joe1/2* (jasmonate overexpressing) mutants were isolated as mutants overexpressing JA-responsive promoter-driven reporter genes (137) and are thought to act upstream of *COI1*. The *cet* (constitutive expression of thionin) mutants were isolated as mutants overexpressing the JA-inducible *Thi2.1* gene and many were found to carry increased levels of JA and its precursor OPDA. These mutants also exhibit enhanced resistance to *Fusarium oxysporum* and spontaneous cell death on their leaves. Interestingly, the *cet2* and *cet9* mutants also appear to be upregulated in the SA pathway, indicating a point of possible overlap between the JA and SA pathways.

The Ethylene Pathway

The perception of ET in Arabidopsis is mediated by five different receptors, namely ETR1, ETR2, ERS1, ERS2 and EIN4. Mutant analysis indicates that all five receptors are required for the induction of subsequent responses to ET (145-148). These receptor proteins share sequence similarities with the bacterial two-component histidine (His) kinases. The NH_4-terminal ends of the five proteins are the most conserved and consist of an ET-binding domain. However the COOH-terminal regions are more divergent. Both the ETR1 and ERS1 proteins (type-I subfamily) carry a conserved His kinase domain, while the His kinase domain of the ETR2, ERS2 and EIN4 (type-II subfamily) proteins is more varied, and lacks residues essential for catalytic activity (149). In addition, the ERS1 and 2 proteins lack a receiver domain, which along with the kinase domain, is involved in interaction between the receptor and CTR1, a downstream negative regulator of the ET pathway. This suggests that neither the receiver domain, nor the His kinase activity, are essential for receptor function. This was further confirmed by the observation that the *etr1 ers1* double mutant can be rescued by overexpression of ETR1 or ERS1, as well as overexpression of a mutant form of etr1 that lacked the His kinase activity (150). Furthermore, a COOH-terminal truncation, as well as site directed mutations, eliminating the His kinase activity in the *etr1-1* mutant continue to confer dominant ET insensitivity to the mutant plants (151). Since the etr1-1 protein lacking His kinase activity is able to actively repress ET responses, it is possible that the NH_4-terminal region of the receptor participates in dimerization with other intact receptors and thus continues to relay downstream signaling in the absence of a functional COOH-terminal domain. Although there is no evidence suggesting heterodimer formation between the different receptors, the ETR1 protein has been shown to form disulfide-linked homodimers (152). Transgenic tobacco plants expressing the mutant *etr1-1* gene are not only ET-insensitive, but also susceptible to opportunistic fungi (153). The *etr1* mutation also affects induced systemic resistance in Arabidopsis plants (154, 155).

Perception of ET possibly occurs in the endoplasmic reticulum (ER), based on the fact that ETR1 (156), and possibly ERS1 and ETR2 (157), are localized to the ER. However, other plant species may perceive ET elsewhere. GFP-tagged transient expression studies of the tobacco NtHKI (histidine kinase-like) receptor, indicate possible localization to the plasma membrane (158). Affinity and specificity of binding to ET is made possible by the hydrophobic ligand-binding pocket of the receptor and a copper co-factor (159), possibly delivered by the copper transporter RAN1. Analysis of loss-of-function mutations in the receptors has demonstrated that binding of ET results in the inactivation of the receptor proteins (148), which in turn, inactivates CTR1 (constitutive triple response), a Raf-like serine/threonine kinase (160). Thus, in the absence of ET, the receptors are active and constitutively induce CTR1. CTR1 has been shown to interact directly with ETR1, ETR2 and ERS2 (161-163) and, although CTR1 lacks a transmembrane domain or membrane attachment motifs, it is localized to the ER (161), possibly *via* its association with the ER-localized receptor protein. While bound to the receptor, CTR1 probably remains in an active conformation and is able to repress downstream ET responses. In the presence of ET, conformational changes in the receptor molecule could in turn result in conformational changes of CTR1, inhibiting its kinase activity. Alternatively, relocation of CTR1 to a site distal from its phosphorylation substrate could result in the derepression of downstream responses. Indeed, interaction of CTR1 with the receptor protein results in the localization of CTR1 to the ER, and this localization is important for the suppression of the ET responses. Similarly, mutations in CTR1 which disrupt its association with the receptors, as well as double- and triple-combination loss-of-function mutations in the receptors, result in redistribution of CTR1 to the cytosol and induce constitutive ET responses (148, 161, 164). The function of CTR1 depends both upon its COOH-terminal Ser/Thr kinase activity and its NH_4-terminal domain-mediated association with the ET receptors. Overexpression of the receptor-associating NH_4-terminal region of CTR1 prevents the binding between the endogenous, functional CTR1 and the ET receptors, resulting in a constitutive ET response phenotype. In contrast, overexpression of the NH_4-terminal region of the ctr1-8 protein, which is disrupted in its association with the receptor, did not constitutively switch on the ET response pathway (164). Moreover, several mutations disrupting the kinase activity of CTR1 result in a constitutive ET response phenotype (164). The CTR1-like MAP kinase kinase kinase (MAPKKK) EDR1 (enhanced disease resistance1), suppresses resistance to powdery mildew causing fungi (165). However, this gene does not appear to participate in the ET pathway and may be involved in regulating SA-related responses.

The CTR1 protein has been shown to interact with the catalytic subunit (PP2A-1C) of PP2A phosphatase (166). The *RCN1* gene encodes the regulatory subunit of this phosphatase complex. A loss-of-function mutation in *rcn1* (167) or mutations affecting the PP2A activity, result in increased sensitivity to ET and exaggeration of the ET responses. However, since CTR1 fails to phosphorylate RCN1 or PP2A-1C, neither of these proteins appear to be substrates for CTR1. Based on the activity of the mammalian Raf proteins, it has been suggested that PP2A reduces CTR1 activity so that lesser amounts of ET are required to induce a response in the plant (166). It is important to remember though, that RCN1 is also known to be involved in auxin signaling (168, 169) and its effect on ET signaling may be the result of cross-talk between the two hormone signaling pathways.

The CTR1 protein shares sequence similarities with the mammalian MAPKKKs, leading to the possibility that a MAPK cascade may be involved in the signal transduction pathway activated by ET (160, 164). This is further supported by the observation that ET stimulates a protein with MAPK activity in Arabidopsis (170). Furthermore, *ctr1* mutants are constitutively activated in MPK6 activity, and the application of the ET-precursor, aminocyclopropane-1-carboxylic acid (ACC) activates specific MAPKs in both Arabidopsis and Medicago (171). These include MPK6 and another MAPK in Arabidopsis, and SIMK and MMK3 in Medicago. Arabidopsis lines overexpressing a SIMKK exhibit the constitutive ET response phenotype similar to the *ctr1* mutant, and constitutively activate MPK6 (171). However, RNA interference (RNAi) and T-DNA insertion analysis of the Arabidopsis *MPK6* does not indicate a role in ET responses (172, 173). Since CTR1 acts as a negative regulator of the ET pathway, it appears that this MAPK cascade would involve the activation of a MAPK *via* the inactivation of a MAPKK by a MAPKKK. This is in contrast to the popularly-accepted concept that MAPKKK activates MAPKK, which in turn, activates MAPK. Thus, although a MAPK cascade may be involved in ET signaling, the actual mechanism by which the ET signal is relayed from CTR1 to EIN2, a downstream, positive regulator of ET signaling, remains to be understood.

In addition to the ET receptors and CTR1, which are all negative regulators of the ET-response pathway, the ET signal transduction cascade also comprises several positive regulators, including EIN2, EIN3, EIN5 and EIN6. A loss-of-function mutation in EIN2 results in complete ET insensitivity (174, 175) and loss of ISR in Arabidopsis (155). Genetic analysis places *EIN2* downstream of *CTR1*. The EIN2 protein consists of an NH_4-terminal domain with sequence similarity to the disease-related Nramp family of ion transporters, which may be involved in sensing the upstream signals. It possesses 12 predicted transmembrane domains, suggesting its involvement in membrane localization. Indeed, biochemical analysis has suggested the membrane association of EIN2 (176). The COOH-terminal region of the protein consists of mostly hydrophilic amino acids, possibly forming a coiled-coil helix, and implying its role in protein-protein interactions. Overexpression of the COOH-end, but not the full-length EIN2 protein, confers constitutive ET responses (176), suggesting that the Nramp-like domain may control the COOH-terminal function of this protein. Although the similarity to Nramp proteins indicates a possible role in ion transport, no such activity has been demonstrated for EIN2. Therefore, the actual function of EIN2 remains to be unraveled.

A second nuclear-localized transcription factor EIN3 functions downstream of EIN2 (177). A mutation in *ein3* renders plants defective in ET-mediated responses. Interestingly, overexpression of this protein induces constitutive ET-responses in both wild-type and *ein2* plants (177), signifying that EIN3 is possibly sufficient for the activation of the ET response pathway. The EIN3 protein is very short-lived and is subject to degradation *via* a ubiquitin/proteosome-dependent pathway, mediated by the F-box proteins EBF1and 2 (EIN3-binding F-box protein). ET inhibits the proteolysis of EIN3 and thereby promotes its accumulation in the nucleus (178). Similarly, the *ebf1* and *ebf2* mutants exhibit ET hypersensitivity and result in increased accumulation of EIN3 in the nucleus. Conversely, overexpression of EBF1 and 2 results in reduced sensitivity to ET. In addition to the EBF proteins, many components of the ET pathway, including the ETR/ERS receptors, EIN2, 5 and 6, are also required for the accumulation of EIN3. Besides *EIN3*, the Arabidopsis genome carries five other *EIN3*-like (*EIL*) genes. However, only the *EIL1* gene appears to have a major role in ET signaling (174). A screen for weak ET insensitive (*wei*) mutants resulted in the identification of *eil1* mutants with increased susceptibility to the necrotrophic fungus *B. cinerea* (179). Overexpression of EIL1, but not of any other EIL proteins (EIL2-5), confers a constitutive ET response phenotype on wild-type as well as *ein2* plants (177). The EIL2, 4 and 5 proteins possibly play minor roles in ET response or may participate in other unrelated pathways.

The EIN3 transcription factor directly induces the ET response factor (ERF1), a GCC-box binding protein (180). EIN3 binds as a dimer, to the primary

ET response element (PERE) in the *ERF1* promoter. Constitutive expression of ERF1 results in enhanced resistance to the necrotrophic fungi, *B. cinerea* and *P. cucumerina* (138), and the soil-borne fungus *Fusarium oxysporum* (181). Overexpression activates several, but not all, the ET response phenotypes, suggesting that ERF1 is responsible for mediating a subset of the ET responses and may act in conjunction with other members of the ET response element binding proteins (EREBPs). In addition to ET, ERF1 can also be induced by JA (182). Mutations in the ET-as well as JA-signaling pathways inhibit the induction of *ERF1* by either hormone. *ERF1* overexpressing lines are restored in their defense-related phenotypes in both the *coi1* and *ein2* backgrounds. Furthermore, the GCC-box required for ERF1 binding in the *PDF1.2* promoter is also known to be responsive to JA (183). Thus, it appears that ERF1 may be a transcription factor involved in integrating signals from the ET and JA pathways and is, thereby, involved in cross-talk between the two pathways. In addition to ERF1, EIN3 possibly modulates the expression of four other members of the EREBP transcription factor family. These genes (EDF1, 2, 3 and 4) carry the B3 DNA-binding domain in addition to AP2, a DNA-binding domain present in all EREBP proteins. Expression of these genes is ET inducible and they affect some of the ET-responsive phenotypes.

ET signaling is also affected by the transcriptional regulation of genes involved in ET biosynthesis. ET itself is involved in the regulation of some of these genes, including the ACC synthase (ACS, 184) and ACC oxidase (ACO, 185) gene families. Some of these in turn are ET-responsive (186), thus entailing a feedback loop mechanism for the control of biosynthesis and response to ET. Others are responsive to multi-hormone signals, such as auxin (187) or gibberillins (188) in conjunction with ET, indicating a role in cross-talk between the different pathways. Post-transcriptional regulation of ET signaling occurs *via* ubiquitin-26S proteosome-mediated degradation. The ACS protein is rapidly degraded upon synthesis (189), and there is evidence that ubiquitination may be involved in the negative regulation of the ACS5 protein (190). In addition, levels of the ACO protein also appear to be under the control of the ubiquitine-mediated degradation process (191). As mentioned previously, the EBF1 and 2 proteins regulate the degradation of EIN3 *via* ubiquitination.

Thus the regulation of the ET signaling pathway involves a complicated relay of signals between multiple components, including cross-talk between several different pathways.

CONCLUSION

It is becoming clear that there is a complex interplay of signal molecules between the various defense-signaling pathways in plants. Furthermore, defense-signaling pathways also appear to employ factors derived from metabolic pathways to fine-tune the defense responses. One example is the Arabidopsis *G6PDH* (glucose-6-phosphate 1-dehydrogenase) gene, which encodes the rate-limiting enzyme of the pentose phosphate pathway, and is induced upon pathogen infection (192). Inhibiting the activity of this enzyme not only affects the redox state of the cell, but also affects SA signaling by preventing the monomerization of

NPR1 and thereby prevents *PR* gene induction (58). Similarly, a glycerol kinase encoded by the *NHOI* (*GLI1*) gene is required for non-host resistance to bacterial and fungal pathogens (193). *NHOI* expression is suppressed in Arabidopsis plants infected with a virulent strain of *Pseudomonas* and this suppression is dependent on an intact JA signaling pathway. Moreover, glycerol and FA metabolism also modulate defense responses in Arabidopsis plants (89, 90). Exogenous application of vitamin B_1 (thiamin), an essential nutritional element, appears to induce SAR, expression of *PR* genes, and resistance to pathogens in a variety of plants, including Arabidopsis (194). The observation that ~25% of the Arabidopsis genes appear to alter their expression in response to pathogen infection (195, 196) implies the existence of a highly sophisticated regulatory mechanism controlling the transcriptional re-programming on a massive scale. Although genetic analysis, examination of T-DNA insertion mutants, and large-scale gene expression studies have greatly enhanced understanding of this defense-related interplay between the different pathways, full comprehension of the mechanisms regulating defense signaling is still a long way off.

ACKNOWLEDGMENTS

We thank David Smith for critical comments on the manuscript and Ludmila Lapchyk for help with editing. Our work is funded by grants from NSF (MCB#0421914), USDA-NRI (2004-03287, 2003-35329-13312) and KSEF (419-RDE-004, 04RDE-006, 820-RDE-007).

REFERENCES

1 Ross, A.F. (1961) Virology 14, 340-358.

2 Pieterse, C.M.J., Van Wees, S.C.M., Ton, J., Van Pelt, J.A. and Van Loon, L.C. (2002) Plant Biol. 4, 535-544.

3 Gundlach, H., Muller, M.J., Kutchan, T.M. and Zenk, M.H. (1992) Proc. Nat. Acad. Sci. U.S.A. 89, 2389-2393.

4 Feys, B., Benedetti, C.E., Penfold, C.N. and Turner, J.G. (1994) Plant Cell 6, 751-759.

5 Menke, F.L., Champion, A., Kijne, J.W. and Memelink, J. (1999) EMBO J. 18, 4455-4463.

6 Brader, G., Tas, E. and Palva, E. T. (2001) Plant Physiol. 126, 849-860.

7 McConn, M. and Browse, J. (1996) Plant Cell 8, 403-416.

8 Mattoo, A.K. and Suttle, J.C. (1991) in The Plant Hormone Ethylene, CRC Press, Inc. Boca Raton, FL.

9 Abeles, F.B., Morgan, P.W. and Saltveit, M.E. (1992) in Ethylene in Plant Biology, Academic Press, San Diego, CA.

10 Schaller, G.E. (2003) in Encyclopedia of Applied Plant Science, (Thomas B, Murphy D, Murray B, eds.), pp. 1019-1026. Academic Press, London, UK.

11 Lamb, C. and Dixon, R.A. (1997) Annu. Rev. Plant Physiol. Plant Mol. Biol. 48, 251-275.

12 Levine, A., Tenhaken, R. Dixon, R.A. and Lamb, C.J. (1994) Cell 79, 583-593.

13 Alvarez, M.E., Pennell, R.I., Meijer, P.J., Ishikawa, A., Dixon, R.A. and Lamb, C.J. (1998) Cell 92, 773-784.
14 Kauss, H. and Jeblick, W. (1995) Plant Physiol. 108, 1171-1178.
15 Shirasu, K., Nakajima, H., Rajasekhar, V.K., Dixon, R.A. and Lamb, C. (1997) Plant Cell 9, 261-270.
16 Mur, L.A., Brown, I.R., Darby, R.M., Bestwick, C.S., Bi, Y.M., Mansfield, J.W. and Draper, J. (2000) Plant J. 23, 609-621.
17 Tierens, K.F., Thomma, B.P., Bari, R.P., Garmier, M., Eggermont, K., Brouwer, M., Penninckx, I.A., Broekaert, W.F. and Cammue, B.P. (2002) Plant J. 29, 131-140.
18 Klessig, D.F., Durner, J., Noad, R., Navarre, D.A., Wendehenne, D., Kumar, D., Zhou, J.M., Shah, J., Zhang, S., Kachroo, P., Trifa, Y., Pontier, D., Lam, E. and Silva, H. (2000) Proc. Nat. Acad. Sci. U.S.A. 97, 8849-8855.
19 Mittler, R., Herr, E.H., Orvar, B.L., van Camp, W., Willekens, H., Inzé, D. and Ellis, B.E. (1999) Proc. Nat. Acad. Sci. U.S.A. 96, 14165-14170.
20 Delledonne, M., Zeier, J., Marocco, A. and Lamb, C. (2001) Proc. Nat. Acad. Sci. U.S.A. 98, 13454-13459.
21 Coego, A., Ramirez, V., Ellul, P., Mayda, E. and Vera, P. (2005) Plant J. 42, 283-293.
22 Coego, A., Ramirez, V., Gil, M.J., Flors, V., Mauch-Mani, B. and Vera, P. (2005) Plant Cell 17, 2123-2137.
23 Malamy, J., Carr, J.P., Klessig, D.F. and Raskin, I. (1990) Science 250, 1002-1004.
24 Uknes, S., Winter, A., Delaney, T., Vernooij, B., Morse, A., Friedrich, L., Nye, G., Potter, S., Ward, E. and Ryals, J. (1993) Mol. Plant-Microbe Interact. 6, 692-698.
25 Dempsey, D., Shah, J. and Klessig, D.F. (1999) Crit. Rev. Plant Sci. 18, 547-575.
26 Dong, X. (2001) Curr. Opin. Plant Biol. 4, 309-314.
27 Glazebrook, J. (2001) Curr. Opin. Plant Biol. 4, 301-308.
28 Kunkel, B.N. and Brooks, D.M. (2002) Curr. Opin. Plant Biol. 5, 325-331.
29 Vernooij, B., Friedrich, L., Morse, A., Reist, R., Kolditz-Jawhar, R., Ward, E., Uknes, S., Kessmann, H. and Ryals, J. (1994) Plant Cell 6, 959-965.
30 Ryals, J. L. Neuenschwander, U.H., Willits, M.G., Molina, A., Steiner, H-Y. and Hunt, M.D. (1996) Plant Cell 8, 1809-1819.
31 Mauch-Mani, B. and Slusarenko, A.J. (1996) Plant Cell 8, 203-212.
32 Verberne, M.C., Budi Muljono, A.B. and Verpoorte, R. (1999) in Biochemistry and Molecular Biology of Plant Hormones (K. Libbenga, M. Hall, P.J.J. Hooykaas, eds.), pp. 295-312. Elsevier, London.
33 Wildermuth, M.C., Dewdney, J., Wu, G. and Ausubel, F.M. (2001) Nature 414, 562-565.
34 Dewdney, J., Reuber, T.L., Wildermuth, M.C., Devoto, A., Cui, J., Stutius, L.M., Drummond, E.P. and Ausubel, F.M. (2000) Plant J. 24, 205-218.
35 Nawrath, C. and Metraux, J.P. (1999) Plant Cell 11, 1393-1404.

36 Nawrath, C., Heck, S., Parinthawong, N. and Métraux, J.-P. (2002) Plant Cell 14, 275-286.
37 Falk, A., Feys, B.J., Frost, L.N., Jones, J.D., Daniels, M.J. and Parker, J.E. (1999) Proc. Nat. Acad. Sci. U.S.A. 96, 3292-3297.
38 Jirage, D., Tootle, T.L., Reuber, T.L., Frost, L.N., Feys, B.J., Parker, J.E., Ausubel, F. M. and Glazebrook, J. (1999) Proc. Nat. Acad. Sci. U.S.A 96, 13583-13588.
39 Shirano, Y., Kachroo, P., Shah, J. and Klessig, D.F. (2002) Plant Cell 14, 3149-3162.
40 Chandra-Shekara, A.C., Navarre, D., Kachroo, A., Kang, H.G., Klessig, D. and Kachroo, P. (2004) Plant J. 40, 647-59.
41 Parker, J.E., Holub, E.B., Frost, L.N., Falk, A., Gunn, N.D. and Daniels, M.J. (1996) Plant Cell 8, 2033-2046.
42 Aarts, N., Metz, M., Holub, E., Staskawicz, B.J., Daniels, M.J. and Parker, J.E. (1998) Proc. Nat. Acad. Sci. U.S.A 95, 10306-10311.
43 Xiao, S., Calis, O., Patrick, E., Zhang, G., Charoenwattana, P., Muskett, P., Parker, J.E. and Turner, J.G. (2005) Plant J. 42, 95-110.
44 Glazebrook, J., Rogers, E.E. and Ausubel, F.M. (1996) Genetics 143, 973-982.
45 Glazebrook, J., Zook, M., Mert, F., Kagan, I., Rogers, E.E., Crute, I.R., Holub, E.B., Hammerschmidt, R. and Ausubel, F.M. (1997) Genetics 146, 381-392.
46 Rustérucci, C., Aviv, D.H., Holt, B.F., Dangl, J.L. and Parker, J.E. (2001) Plant Cell 13, 2211-2224.
47 Brodersen, P., Petersen, M., Pike, H.M., Olszak, B., Skov, S., Odum, N., Jorgensen, L.B., Brown, R.E. and Mundy, J. (2002) Genes Dev. 16, 490-502.
48 Mateo, A., Muhlenbock, P., Rustérucci, C., Chang, C.C., Miszalski, Z., Karpinska, B., Parker, J.E., Mullineaux, P.M. and Karpinski, S. (2004) Plant Physiol. 136, 2818-2830.
49 Feys, B.J., Moisan, L.J., Newman, M.A. and Parker, J.E. (2001) EMBO J. 20, 5400-5411.
50 Feys, B.J., Wiermer, M., Bhat, R.A., Moisan, L.J., Medina-Escobar, N., Neu, C., Cabral, A. and Parker, J.E. (2005) Plant Cell 17, 2601-2613.
51 Century, K.S., Shapiro, A.D., Repetti, P.P., Dahlbeck, D., Holub, E. and Staskawicz, B.J. (1997) Science 278, 1963-1965.
52 Century, K.S., Holub, E.B. and Staskawicz, B.J. (1995) Proc. Nat. Acad. Sci. U.S.A. 92, 6597-6601.
53 Shapiro, A.D. and Zhang, C. (2001) Plant Physiol. 127, 1089-1101.
54 Delaney, T.P., Friedrich, L. and Ryals, J.A., (1995) Proc. Nat. Acad. Sci. U.S.A. 92, 6602-6606.
55 Shah, J., Tsui, F. and Klessig, D.F., (1997) Mol. Plant-Microbe Interact. 10, 69-78.
56 Cao, H., Bowling, S.A., Gordon, S. and Dong, X., (1994) Plant Cell 6, 1583-1592.
57 Cao, H., Glazebrook, J., Clark, J.D., Volko, S. and Dong, X., (1997) Cell 88, 57-64.

58 Mou, Z., Fan, W. and Dong, X. (2003) Cell 113, 935-944.
59 Zhang, Y., Fan, W., Kinkema, M., Li, X. and Dong, X. (1999) Proc. Nat. Acad. Sci. U.S.A. 96, 6523-6528.
60 Zhou, J.-M., Trifa, Y., Silva, H., Pontier, D., Lam, E., Shah, J. and Klessig, D.F. (2000) Mol. Plant-Microbe Interact. 13, 191-202.
61 Després, C., DeLong, C., Glaze, S., Liu, E. and Fobert, P.R. (2000) Plant Cell 12, 279-290.
62 Kim, H.S. and Delaney, T.P. (2002) Plant J. 32, 151-163.
63 Després, C., Chubak, C., Rochon, A, Clark, R., Bethune, T., Desveaux, D. and Fobert, P.R. (2003) Plant Cell 15, 2181-2191.
64 Zhang, Y., Tessaro, M.J., Lassner, M. and Li, X. (2003) Plant Cell 15, 2647-2653.
65 Felton, G.W. and Korth, K.L. (2000) Curr. Opin. Plant Biol. 3, 309-314.
66 Feys, B.J. and Parker, J.E. (2000) Trends Genet. 16, 449-455.
67 Pieterse, C.M.J., Ton, J. and Van Loon, L.C. (2001) AgBiotechNet 3, ABN 068.
68 Spoel, S.H., Koornneef, A., Claessens, S.M.C., Korzelius, J.P., Van Pelt, J.A., Mueller, M.J., Buchala, A.J., Métraux,J.-P., Brown, R., Kazan, K., Van Loon, L.C., Dong, X. and Pieterse, C.M.J. (2003) Plant Cell 15, 760-770.
69 Thomma, B.P.H.J., Penninckx, I.A.M.A., Cammue, B.P.A., and Broekaert, W.F. (2001). Curr. Opin. Immunol. 13, 63-68.
70 van Wees, S.C., de Swart, E.A., van Pelt, J.A., van Loon, L.C. and Pieterse, C.M. (2000) Proc. Nat. Acad. Sci. U.S.A. 97, 8711-8716.
71 Iavicoli, A., Boutet, E., Buchala, A. and Metraux, J.P. (2003) Mol. Plant Microbe Interact. 16, 851-858.
72 Li, X., Zhang, Y., Clarke, J.D., Li, Y. and Dong, X. (1999) Cell 98, 329-339.
73 Weigel, R.R., Pfitzner, U.M. and Gatz, C. (2005) Plant Cell 17, 1279-1291.
74 Li, J., Brader, G. and Palva, E.T. (2004) Plant Cell 16, 319-331.
75 Desveaux, D., Subramaniam, R., Despres, C., Mess, J.N., Levesque, C., Fobert, P.R., Dangl, J.L. and Brisson, N. (2004) Dev. Cell 6, 229-240.
76 Rate, D.N., Cuenca, J.V., Bowman, G.R., Guttman, D.S. and Greenberg, J.T. (1999) Plant Cell 11, 1695-1708.
77 Clarke, J.D., Volko, S.M., Ledford, H., Ausubel, F.M. and Dong, X. (2000) Plant Cell 12, 2175-2190.
78 Weymann, K., Hunt, M., Uknes, S., Neuenschwander, U., Lawton, K., Steiner, H. Y. and Ryals, J. (1995) Plant Cell 7, 2013-2022.
79 Aviv, D.H., Rustérucci, C., Holt, III, B.F., Dietrich, R.A., Parker, J.E. and Dangl, J.L. (2002) Plant J. 29, 381-391.
80 Yu, I.C., Parker, J. and Bent, A.F. (1998) Proc. Nat. Acad. Sci. U.S.A. 95, 7819-7824.
81 Rate, D.N. and Greenberg, J.T. (2001) Plant J. 27, 203-211.
82 Kachroo, P., Shanklin, J., Shah, J., Whittle, E.J. and Klessig, D.F. (2001) Proc. Nat. Acad. Sci. U.S.A. 98, 9448-9453.
83 Dietrich, R.A., Richberg, M.H., Schmidt, R., Dean, C. and Dangl, J.L. (1997) Cell 88, 685-694.

84 Lu, H., Rate, D.N., Song, J.T. and Greenberg, J.T. (2003) Plant Cell 15, 2408-2420.
85 Petersen, M., Brodersen, P., Naested, H., Andreasson, E., Lindhart, U., Johansen, B., Nielsen, H.B., Lacy, M., Austin, M.J., Parker, J.E., Sharma, S.B., Klessig D.F., Martienssen, R., Mattsson, O., Jensen, A.B. and Mundy, J. (2000) Cell 103, 1111-1120.
86 Andreasson, E., Jenkins, T., Brodersen, P., Thorgrimsen, S., Petersen, N.H., Zhu, S., Qiu, J.L., Micheelsen, P., Rocher, A, Petersen, M., Newman, M.A., Bjorn Nielsen, H., Hirt, H., Somssich, I., Mattsson, O. and Mundy, J. (2005) EMBO J. 24, 2579-2589.
87 Durrant, W.E., Rowland, O., Piedras, P., Hammond-Kosack, K.E. and Jones, J.D. (2000) Plant Cell 12, 963-977.
88 Kachroo, P., Kachroo, A., Lapchyk, L., Hildebrand, D. and Klessig, D.F. (2003) Mol. Plant Microbe-Interact. 16, 1022-1029.
89 Kachroo, A., Lapchyk, L., Fukushige, H., Hildebrand, D., Klessig, D. and Kachroo, P. (2003) Plant Cell. 15, 2952-2965.
90 Kachroo, A., Venugopal, S.C., Lapchyk, L., Falcone, D., Hildebrand, D. and Kachroo, P. (2004) Proc. Nat. Acad. Sci. U.S.A. 101, 5152-5157.
91 Nandi, A., Welti, R. and Shah, J. (2004) Plant Cell 16, 465-477.
92 Maldonado, A.M., Doerner, P., Dixon, R.A., Lamb, C.J. and Cameron, R.K. (2002) Nature 419, 399-403.
93 Kumar, D. and Klessig, D.F. (2003) Proc. Nat. Acad. Sci. U.S.A. 100, 16101-16106.
94 Chen, Z., Silva, H. and Klessig, D.F. (1993) Science 262, 1883-1886.
95 Slaymaker, D.H., Navarre, D.A., Clark, D., del Pozo, O., Martin, G.B. and Klessig, D.F. (2002) Proc. Nat. Acad. Sci. U.S.A. 99, 11640-11645.
96 Durner, J. and Klessig, D.F. (1995) Proc. Nat. Acad. Sci. U.S.A. 92, 11312-11316.
97 Neuenschwander, U., Vernooij, B., Friedrich, L., Uknes, S., Kessmann, H. and Ryals, J. (1995) Plant J. 8, 227-233.
98 Bi, Y.M., Kenton, P., Mur, L., Darby, R. and Draper, J. (1995) Plant J. 8, 235-245.
99 Forouhar, F., Yang, Y., Kumar, D., Chen, Y., Fridman, E., Park, S.W., Chiang, Y., Acton, T.B., Montelione, G.T., Pichersky, E., Klessig, D.F. and Tong, L. (2005) Proc. Nat. Acad. Sci. U.S.A. 102, 1773-1778.
100 Turner, J.G., Ellis, C. and Devoto, A. (2002) Plant Cell 14, S153-164.
101 Devoto, A. and Turner, J.G. (2003) Ann. Bot. (Lond). 92, 329-337.
102 McConn, M., Creelman, R.A., Bell, E., Mullet, J.E. and Browse, J. (1997) Proc. Nat. Acad Sci. U.S.A. 94, 5473-5477.
103 Vijayan, P., Shockey, J., Levesque, C.A., Cook, R.J. and Browse, J. (1998) Proc. Nat. Acad. Sci. U.S.A. 95, 7209-7214.
104 Howe, G.A., Lightner, J., Browse, J. and Ryan, C.A. (1996) Plant Cell 8, 2067-2077.
105 Constabel, C.P., Bergey, D.R. and Ryan, C.A. (1995) Proc. Nat. Acad. Sci. U.S.A. 92, 407-411.
106 Pearce, G., Strydom, D., Johnson, S. and Ryan, C.A. (1991) Science 253, 895-898.

107 McGurl, B., Pearce, G., Orozco-Cardenas, M. and Ryan, C.A. (1992) Science 255, 1570-1573.
108 Orozco-Cárdenas, M., McGurl, B. and Ryan, C.A. (1993) Proc. Nat. Acad. Sci. U.S.A. 90, 8273-8276.
109 McGurl, B., Orozco-Cardenas, M., Pearce, G. and Ryan, C.A. (1994) Proc. Nat. Acad. Sci. U.S.A. 91, 9799-9802.
110 Li, L., Li, C., Lee, I.G. and Howe, G.A. (2002) Proc. Nat. Acad. Sci. U.S.A. 99, 6416-6421.
111 Meindl, T., Boller, T. and Felix, G. (1998) Plant Cell 10, 1561-1570.
112 Scheer, J.M. and Ryan, C.A. (1999) Plant Cell 11, 1525-1536.
113 Scheer, J.M. and Ryan, C.A. Jr. (2002) Proc. Nat. Acad. Sci. U.S.A. 99, 9585-9590.
114 Farmer, E.E. and Ryan, C.A. (1992) Plant Cell 4, 129-134.
115 Felix, G. and Boller, T. (1995) Plant J. 7, 381-389.
116 Moyen, C. and Johannes, E. (1996) Plant Cell Environ. 19, 464-470.
117 Schaller, A. and Oecking, C. (1999) Plant Cell 11, 263-272.
118 Stratmann, J.W. and Ryan, C.A. (1997) Proc. Nat. Acad. Sci. U.S.A. 94, 11085-11089.
119 Narváez-Vásquez, J., Florin-Christensen, J. and Ryan, C.A. (1999) Plant Cell 11, 2249-2260.
120 Li, C., Williams, M.M., Loh, Y.T., Lee, G.I. and Howe, G.A. (2002) Plant Physiol. 130, 494-503.
121 Li, C., Liu, G., Xu, C., Lee, G.I., Bauer, P., Ling, H.Q., Ganal, M.W. and Howe, G.A. (2003) Plant Cell 15, 1646-1661.
122 Benedetti, C.E., Xie, D. and Turner, J.G. (1995) Plant Physiol. 109, 567-572.
123 Xie, D.X., Feys, B.F., James, S., Nieto-Rostro, M. and Turner, J.G. (1998) Science 280, 1091-1094.
124 Penninckx, I.A., Thomma, B.P., Buchala, A., Metraux J.P. and Broekaert, W.F. (1998) Plant Cell 10, 2103-2113.
125 Thomma, B.P., Eggermont, K., Tierens, K.F. and Broekaert, W.F. (1999) Plant Physiol. 121, 1093-1102.
126 Thomma, B., Eggermont, K., Penninckx, I., MauchMani, B., Vogelsang, R., Cammue, B.P.A. and Broekaert, W.F. (1998) Proc. Nat. Acad. Sci. U.S.A. 95, 15107-15111.
127 Bai, C., Sen, P., Hofmann, K., Ma, L., Goebl, M., Harper, J.W. and Elledge, S.J. (1996) Cell 86, 263-274.
128 Xu, L., Liu, F., Lechner, E., Genschik, P., Crosby, W.L., Ma, H., Peng, W., Huang, D. and Xie D. (2002) Plant Cell 14, 1919-1935.
129 Devoto, A., Nieto-Rostro, M., Xie, D., Ellis, C., Harmston, R., Patrick, E., Davis, J., Sherratt, L., Coleman, M. and Turner, J.G. (2002) Plant J. 32, 457-466.
130 Seigneurin-Berny, D., Verdel, A., Curtet, S., Lemercier, C., Garin, J., Rousseaux, S. and Khochbin, S. (2001) Mol. Cell Biol. 21, 8035-8044.
131 Xiao, S., Dai, L., Liu, F., Wang, Z., Peng, W. and Xie, D. (2004) Plant Cell 16, 1132-1142.
132 Staswick, P.E., Yuen, G.Y. and Lehman, C.C. (1998) Plant J. 15, 747-754.

133 Black, P.N., DiRusso, C.C., Sherin, D., MacColl, R., Knudsen, J. and Weimar, J. D. (2000) J. Biol. Chem. 275, 38547-38553.
134 Staswick, P.E., Tiryaki, I. and Rowe, M.L. (2002) Plant Cell 14, 1405-1415.
135 Staswick, P.E. and Tiryaki, I. (2004) Plant Cell 16, 2117-2127.
136 Berger, S., Bell, E. and Mullet, J.E. (1996) Plant Physiol. 111, 525-531.
137 Jensen, A.B., Raventos, D. and Mundy, J. (2002) Plant J. 29, 595-606.
138 Lorenzo, O., Chico, J.M., Sanchez-Serrano, J.J. and Solano, R. (2004) Plant Cell 16, 1938-1950.
139 Ichimura, K., Mizoguchi, T., Yoshida, R., Yausa, T. and Shinozaki, K. (2000) Plant J. 24, 655-665.
140 Penninckx, I.A.M.A., Eggermont, K., Terras, F.R.G., Thomma, B.P.H.J., DeSamblanx, G.W., Buchala, A., Metraux, J.-P., Manners, J.M., and Broekaert, W.F. (1996) Plant Cell 8, 2309-2323.
141 Epple, P., Apel, K. and Bohlmann, H. (1995) Plant Physiol. 109, 813-820.
142 Ellis, C. and Turner, J.G. (2001) Plant Cell 13, 1025-1033.
143 Hilpert, B., Bohlmann, H., op den Camp, R.O., Przybyla, D., Miersch, O., Buchala, A. and Apel, K. (2001) Plant J. 26, 435-446.
144 Xu, L., Liu, F., Wang, Z., Peng, W., Huang, R., Huang, D. and Xie, D. (2001) FEBS Lett. 494, 161-164.
145 Chang, C., Kwok, S.F., Bleecker, A.B. and Meyerowitz, E.M. (1993) Science 262, 539-544.
146 Hua, J., Chang, C., Sun, Q. and Meyerowitz, E.M. (1995) Science 269, 1712-1714.
147 Sakai, H., Hua, J., Chen, Q.G., Chang, C., Medrano, L.J., Bleecker, A.B. and Meyerowitz, E.M. (1998) Proc. Nat. Acad. Sci. U.S.A. 95, 5812-5817.
148 Hua, J. and Meyerowitz, E.M. (1998) Cell 94, 261-271.
149 Chang, C. and Shockey, J.A. (1999) Curr. Opin. Plant Biol. 2, 352-358.
150 Wang, W., Hall, A.E., O'Malley, R. and Bleecker, A.B. (2003) Proc. Nat. Acad. Sci. U.S.A. 100, 352-357.
151 Gamble, R.L., Qu, X. and Schaller, G.E. (2002) Plant Physiol. 128, 1428-1438.
152 Schaller, G.E., Ladd, A.N., Lanahan, M.B., Spanbauer, J.M. and Bleecker, A.B. (1995) J. Biol. Chem. 270, 12526-12530.
153 Knoester, M., van Loon, L.C., van den Heuvel, J., Hennig, J., Bol, J.F. and Linthorst, H.J.M. (1998) Proc. Nat. Acad. Sci. U.S.A. 95,1933-1937.
154 Pieterse, C.M., van Wees, S.C., van Pelt, J.A., Knoester, M., Laan, R., Gerrits. H., Weisbeek, P.J. and van Loon, L.C. (1998) Plant Cell 10, 1571-1580.
155 Knoester, M., Pieterse, C.M., Bol, J.F. and Van Loon, L.C. (1999) Mol. Plant Microbe Interact. 12, 720-727.
156 Chen, Y.F., Randlett, M.D., Findell J.I. and Schaller G.E. (2002) J. Biol. Chem. 277, 19861-19866.
157 Chen, Y.F., Etheridge, N. and Schaller, G.E. (2005) Annals Bot. 95, 901-915.

158 Xie, C., Zhang, J.S., Zhou, H.L., Li, J., Zhang, Z.G., Wang, D.W. and Chen, S.Y. (2003) Plant J. 33, 385-393.
159 Rodriguez, F.I., Esch, J.J., Hall, A.E., Binder, B.M., Schaller, G.E. and Bleecker, A.B. (1999) Science 283, 996-998.
160 Keiber, J.J., Rothenberg, M., Roman, G., Feldmann, K.A. and Ecker, J.R. (1993) Cell 72, 427-441.
161 Gao, Z., Chen, Y.F., Randlett, M.D., Zhao, X.C., Findell, J.L., Kieber, J.J. and Schaller, G.E. (2003) J. Biol. Chem. 278, 34725-34732.
162 Clark, K.L., Larsen, P.B., Wang, X. and Chang, C. (1998) Proc. Nat. Acad. Sci. U.S.A. 95, 5401-5406.
163 Cancel, J.D. and Larsen, P.B. (2002) Plant Physiol. 129, 1557-1567.
164 Huang, Y., Li, H., Hutchison, C.E., Laskey, J. and Kieber, J.J. (2003) Plant J. 33, 221-233.
165 Frye, C.A., Tang, D. and Innes, R.W. (2001) Proc. Nat. Acad. Sci. U.S.A. 98, 373-378.
166 Larsen, P.B. and Cancel, J.D. (2003) Plant J. 34, 709-718.
167 Larsen, P.B. and Chang, C. (2001) Plant Physiol. 125, 1061-1073.
168 Garbers, C., DeLong, A., Deruere, J., Bernascomi, P. and Soll, D. (1996) EMBO J. 15, 2115-2124.
169 Rashotte, A.M., DeLong, A. and Muday, G.K. (2001) Plant Cell 13, 1683-1697.
170 Novikova, G.V., Moshkov, I.E., Smith, A.R. and Hall, M.A. (2000) FEBS Lett. 474, 29-32.
171 Ouaked, F., Rozhon, W., Lecourieux, D. and Hirt, H. (2003) EMBO J. 22, 1282-1288.
172 Ecker, J.R. (2004) Plant Cell 16, 3169-3173.
173 Menke, F.L.H., van Pelt, J.A., Pieterse, C.M.J. and Klessig, D.F. (2004) Plant Cell 16, 897-907.
174 Roman, G., Lubarsky, B., Kieber, J.J., Rothenberg, M. and Ecker, J.R. (1995) Genetics 139, 1393-1409.
175 Chen, Q. and Bleecker, A. (1995) Plant Physiol. 108, 597-607.
176 Alonso, J.M., Hirayama, T., Roman, G., Nourizadeh, S. and Ecker, J.R. (1999) Science 284, 2148-2152.
177 Chao, Q., Rothenberg, M., Solano, R., Roman, G., Terzaghi, W. and Ecker, J.R. (1997) Cell 89, 1133-1144.
178 Guo, H. and Ecker, J.R. (2003) Cell 115, 667-677.
179 Alonso, J.M., Stepanova, A.N., Solano, R., Wisman, E., Ferrari, S., Ausubel, F.M. and Ecker, J.R. (2003) Proc. Nat. Acad. Sci. U.S.A. 100, 2992-2997.
180 Solano, R., Stepanova, A., Chao, Q. and Ecker, J.R. (1998) Genes Dev. 12, 3703-3714.
181 Berrocal-Lobo, M. and Molina, A. (2004) Mol. Plant Microbe Interact. 17, 763-770.
182 Lorenzo, O., Piqueras, R., Sanchez-Serrano, J.J. and Solano, R. (2003) Plant Cell 15, 165-178.
183 Brown, R.L., Kazan, K., McGrath, K.C., Maclean D.J. and Manners, J.M. (2003) Plant Physiol. 132, 1020-1032.

184 Liang, X., Oono, Y., Shen, N.F., Kohler, C., Li, K., Scolnik, P.A. and Theologis, A. (1995) Gene 167, 17-24.
185 Prescott, A.G. and John, P. (1996) Annu. Rev. Plant Physiol. Plant Mol. Biol. 47, 245-271.
186 Alonso, J.M., Stepanova, A.N., Leisse, T.J., Kim, C.J., Chen, H., Shinn, P., Stevenson, D.K., Zimmerman, J., Barajas, P., Cheuk, R., Gadrinab, C., Heller, C., Jeske, A., Koesema, E., Meyers, C.C., Parker, H., Prednis, L., Ansari, Y., Choy, N., Deen, H., Geralt, M., Hazari, N., Hom, E., Karnes, M., Mulholland, C., Ndubaku, R., Schmidt, I., Guzman, P., Aguilar-Henonin, L., Schmid, M., Weigel, D., Carter, D.E., Marchand, T., Risseeuw, E., Brogden, D., Zeko, A., Crosby, W.L., Berry, C.C. and Ecker, J.R. (2003) Science 301, 653-657.
187 Ecker, J.R. (1995) Science 268, 667-675.
188 Vierzen, W.H., Achard, P., Harberd, N.P. and Van Der Straeten, D. (2004) Plant J. 37, 505-516.
189 Kim, W.T. and Yang, S.F. (1994) Planta 194, 223-229.
190 Wang, K.L., Yoshida, H., Lurin, C. and Ecker, J.R. (2004) Nature 428, 945-950.
191 Larsen, P.B. and Cancel, J.D. (2004) Plant J. 38, 626-638.
192 Dong, X. (2004) Curr. Opin. Plant Biol. 7, 547-552.
193 Kang, L., Li, J., Zhao, T., Xiao, F., Tang, X., Thilmony, R., He, S. and Zhou, J.M. (2003) Proc. Nat. Acad. Sci. U.S.A. 100, 3519-3524.
194 Ahn, I.-P., Kim, S. and Lee, Y.-H. (2005) Plant Physiol. 138, 1505-1515.
195 Maleck, K., Levine, A., Eulgem, T., Morgan, A., Schmid, J., Lawton, K.A., Dangl, J.L. and Dietrich, R.A. (2000) Nat. Genet. 26, 403-410.
196 Tao, Y., Xie, Z., Chen, W., Glazebrook, J., Chang, H.S., Han, B., Zhu, T., Zou, G. and Katagiri, F. (2003) Plant Cell 15, 317-330.

PROXIMITY LIGATION: A SPECIFIC AND VERSATILE TOOL FOR THE PROTEOMIC ERA

Ola Söderberg, Karl-Johan Leuchowius, Masood Kamali-Moghaddam, Malin Jarvius, Sigrun Gustafsdottir, Edith Schallmeiner, Mats Gullberg, Jonas Jarvius and Ulf Landegren

University of Uppsala
Department of Genetics and Pathology
Rudbeck laboratory
SE-751 85 Uppsala
Sweden

ABSTRACT

Knowledge about the total human genome sequence now provides opportunities to study its myriad gene products. However, the presence of alternative splicing, post-translational modifications, and innumerable protein-protein interactions among proteins occurring at widely different concentrations, all combine to place extreme demands on the specificity and sensitivity of assays. The choice of method also depends on matters such as whether proteins will be analyzed in body fluids and lysates, or localized inside single cells. In this review we discuss commonly used detection methods and compare these to the recently-developed proximity ligation technique.

INTRODUCTION

The problem can be illustrated by the counting of swans in a pond inhabited by swans (*Cygnus olor*) and ducks (*Anas platyrhynchos*). To discriminate

Genetic Engineering, Volume 28, Edited by J. K. Setlow

between the two species we set up the following criteria characterizing a swan: it is white, and it has a long neck. Either one of these criteria may suffice to determine if a bird in the pond is a swan or not. However, in a more complex environment such as a lake the analysis may result in an overestimate of the number of swans. Seagulls (*Larus canus*) are white birds and fulfill the first criterion. Similarly, the second criterion (long neck) is also applicable to herons (*Ardea cinerea*). Specificity of detection increases greatly, however, if the two criteria are combined.

Turning now to proteomics, antibodies do not recognize whole proteins, but merely epitopes on the proteins, each composed of just a few amino acids. Assays that require positive identification of target proteins by two antibodies specific for different epitopes on the same protein exhibit profoundly enhanced specificity over assays that depend on single-binding events. The strategy of double recognition has in fact been used for almost forty years in sandwich immunoassays for soluble proteins, where one immobilized antibody traps the protein that is then detected by a second, labeled, antibody (1, 2). Such assays can reach pM detection levels.

Antibody-based measurements of target protein concentrations generally involve detection of bound antibodies *via* labels such as heavy metals, radioisotopes, fluorophores (3), or using chemoluminescence or linked enzymes (4). Enzyme-linked detection reactions offer enhanced sensitivity because of the catalytic activity of the enzymes, generating detectable products. It is possible to further enhance detection by attaching strands of DNA to antibodies. Such DNA strands can be exponentially amplified by methods like polymerase chain reaction (PCR) (immuno-PCR) (5), or they can be used to prime a rolling circle amplification (RCA) reaction for localized signal amplification by generating a long concatemeric product, covalently linked to the antibody, to which labeled probes can hybridize (immuno-RCA) (6-8). All these means of detecting bound antibodies fail to distinguish among specifically and nonspecifically bound antibodies, however, limiting the increase of signals over background that can be attained, and thus the assay sensitivity.

Proteins are being analyzed in many different formats including simplex assays of proteins in solution, *in situ* analyses and using protein arrays. Protein microarrays (reviewed by Espina et al., 9) allow simultaneous detection of several different proteins. In one format all proteins in a sample to be analyzed are directly labeled, and then the ones that have bound to antibodies immobilized in an array can be detected after washes. The specificity and thereby sensitivity of such assays are inherently limited by the fact that single-binding-events per target molecule are scored. Alternatively, different immobilized bait molecules capture their cognate proteins to specific locations on an array. Subsequently, the bound proteins can be detected and quantified by binding of specific-labeled-antibodies in a sandwich format. The requirement for dual binding increases specificity, but the parallel assay format creates increased opportunities for crossreactive binding between noncognate pairs of protein-binding reagents compared to single-analyte assays, limiting-assay sensitivity and ultimately the degree of multiplexing that can be achieved. The use of arrays of photoaptamers can provide increased specificity without a concomitant increase in risks of crossreactivity with increasing numbers of analytes (10). After excess proteins have been removed from arrays of immobilized photoaptamers by washes, specifically-bound proteins can be covalently cross-linked to the aptamers by treatment with UV light. The requirement for the proteins to be correctly positioned for cross-linking introduces an additional level of specificity and allows nonspecifically bound proteins to be removed by extremely stringent washes, leaving only covalently-bound proteins, followed by detection *via* general protein stains.

Western blots achieve increased specificity of detection by distinguishing proteins both according to electrophoretic mobility and by antibody binding to the separated proteins blotted onto a membrane, resulting in nM detection limits. Localized protein detection reactions are also used in immunohistochemistry, where the distribution of proteins in tissues and cells is revealed by the binding of specific-labeled-antibodies. This information represents an invaluable resource in research and diagnostics. *In situ* assays provide increased information over solution-phase quantification assays since heterogeneity among cells can be revealed and spatial relationships among structures are visualized. All current localized detection reactions suffer from the problem that visualization depends on single binding events, however, limiting specificity and sensitivity. It is also a problem that evaluation of *in situ* staining reactions is subjective, and limited to a relatively crude estimation of the degree of staining. Simultaneous phenotyping of multiple cell types can be performed in solution by flow cytometry determining expression levels of several proteins on the surfaces of individual cells. The protein expression pattern in combination with the light scattering properties of the cells enables highly accurate identification of cell types and maturation stages, with the additional advantage that cells also can be sorted for subsequent analysis.

Despite the multitude of formats for protein detection, there clearly remains a pressing need for highly specific detection reactions in order to negotiate protein concentration ranges that may exceed 10 orders of magnitude in serum, and to allow detection of even single proteins in cells and evaluate the company they keep. The choice of method also depends on matters such as whether proteins will be analyzed in body fluids and lysates, or localized inside single cells. In this review we discuss commonly-used detection methods and compare these to the recently-developed proximity ligation technique

PROXIMITY LIGATION

Recently a new and quite general approach to protein detection—proximity ligation—was described. In this technique specific and sensitive detection can be achieved by utilizing a combination of highly-specific target recognition and powerful signal-amplification as required for detection of low abundant proteins (11, 12). The probes used in proximity ligation are composed of an antigen binding part (e.g., an antibody or an aptamer) to which short single-stranded DNA molecules have been conjugated. Upon binding of two such proximity probes to the same target molecule, a subsequently added connector oligonucleotide can hybridize to the ends of the conjugated DNA strands and guide their joining by enzymatic ligation. This creates a DNA molecule that can then be amplified by PCR (Figure 1). Recognition of target molecules by proximity

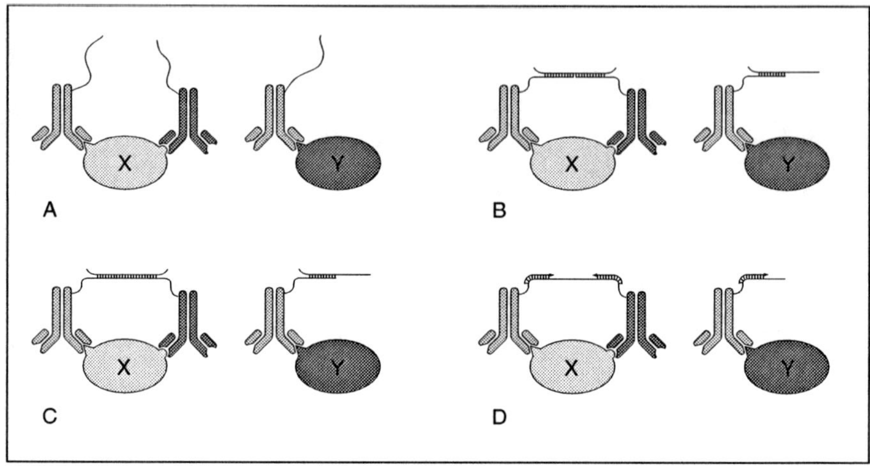

Figure 1. (A) The proximity ligation procedure. Two proximity probes bind protein X, while one probe also crossreacts by binding to protein Y. (B) A complementary connector oligonucleotide is added that hybridizes to the oligonucleotides attached to pairs of adjacent proximity probes, allowing the free oligonucleotide ends to be joined by ligation. (C) Only reagents brought in proximity by binding pair-wise to protein X will be ligated together. (D) Addition of PCR primers allows sensitive detection by exponential amplification of ligated proximity probes having bound protein X, but not of unreacted proximity probes.

ligation thus strictly depends on dual recognition in order to generate an amplifiable DNA strand that serves as a surrogate marker for the detected protein. Signal amplification by real-time PCR allows sub-pM levels of proteins to be detected in a homogenous assay that is performed without any washes, just the addition of a ligation/amplification cocktail, followed by amplification and detection (Figure 2). Alternatively, a sandwich format can be used, where the target proteins are first trapped on a solid support *via* specific binding followed by addition of pairs of proximity probes that are joined by ligation. The removal of excess reagents by washes lowers the background from chance proximity by unbound proximity probes and also reduces the concentration of substances that may inhibit ligation, amplification or detection. The assay involves three recognition events of any target molecule, further increasing the ability to discriminate among closely similar protein molecules.

By virtue of the presence of many copies of the same proteins on their surfaces, even single viral particles or bacteria have been successfully detected using the proximity ligation mechanism (13). It is also possible to design homogenous assays that require three recognition events and two ligations. As a consequence of the reduced chance for proximity of three rather than two reagents, and the increased biological specificity of the three binding events, detection levels of just a few hundred molecules have been achieved (Schallmeiner et al., submitted).

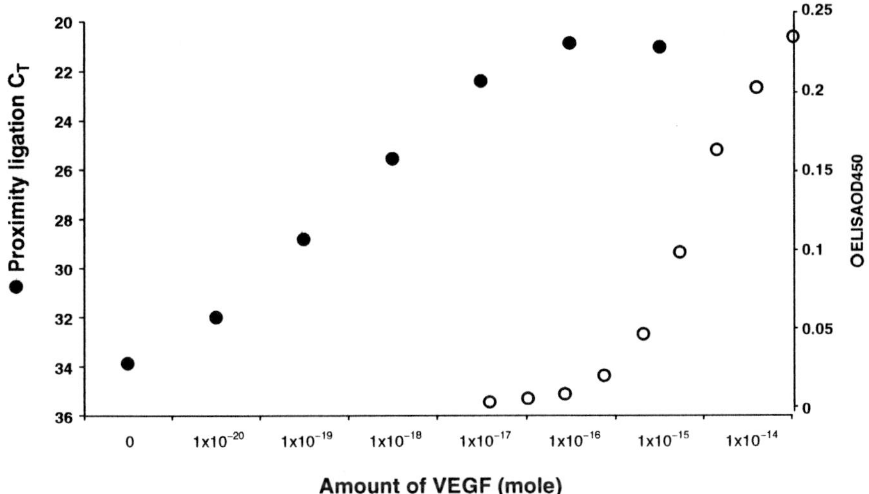

Figure 2. Comparison of detection of VEGF by proximity ligation (filled circles) and by ELISA (open circles). The molar amount of target protein present in 1 μl samples for proximity ligation and 100 μl samples for ELISA is plotted against the cycle threshold values from real-time PCR assays or absorbance at 450 nm for the ELISA

PROXIMITY-LIGATION *IN SITU* ASSAY (P-LISA)

The proximity ligation mechanism can also be used to achieve dual-recognition *in situ* immuno-staining, by modifying the method to provide localized detection signals. In order to obtain highly specific detection *in situ*, the creation of a circular amplifiable DNA template was made dependent on the proximal binding of two proximity probes, in analogy to padlock probe-based detection of single target DNA sequences *in situ* (14). In both cases—using padlock probes for DNA detection and proximity ligation to detect proteins—circular DNA strands form upon highly-specific target detection, and next give rise to single-stranded amplification products composed of hundreds of complements of the circular DNA strands, anchored at the site of probe binding. The RCA products bundle up in random coils less than a micrometer in diameter to which fluorophore-labeled oligonucleotide probes are hybridized. Even single molecules can thus be detected and enumerated easily in a standard fluorescence microscope either by the investigator or using dedicated software, increasing throughput and objectivity (15). Compared with previous methods the requirement for dual recognition significantly increases the specificity of detection.

CURRENT METHODS TO DETECT PROTEIN INTERACTIONS

Measurement of expression levels of a protein often is not sufficient to determine its activity state. Interaction with partners in the formation of protein complexes, and post-translational modifications such as phosphorylation, are often crucial for the functionality of a protein. Post-translational modifications can be studied with specific antibodies binding the modified residues, although as usual crossreactivity remains a problem in single-recognition strategies. An even more difficult task is studies of interactions between proteins, as microscopic co-localization of signals offers too poor resolution to determine if two or more proteins are interacting, due to the limiting resolution and sensitivity of light microscopy. Recent improvements in confocal microscopy such as 4Pi and STED have enhanced the resolution down to 28 nm (see review by Hell (16)), but detection reactions still face problems of crossreactivity and poor detectability of single fluorophores. If information about sub-cellular localization or inter-cellular variation is not required, then gel electrophoresis-based methods such as co-immunoprecipitation are applicable for studies of protein interaction.

In recent years several methods have been developed for detecting protein interactions based upon split-enzymes, where one part of an enzyme is fused with one protein and the other part of the enzyme with a possible interaction partner (17-21). In yeast two-hybrid assays the DNA binding domain of a transcription factor is fused to one protein and the transcription activating domain fused to another protein, restoring the function of the transcription factor only if the two fused proteins interact (17, 18). By using yeast two-hybrid or split ubiquitin (21) whole protein interaction networks can be determined.

Techniques utilizing either split fluorescent/bioluminescent proteins (22, 23) or resonance energy transfer (24-27) have been developed during the last few years for visualization of protein interactions in living cells. In bimolecular fluorescence complementation (BiFC) analysis, the gene encoding yellow fluorescent protein (YFP) is split in two parts and fused with genes for proteins whose interactions are to be monitored (22). Upon interaction between the two fusion proteins, the two halves of YFP are brought together, resulting in fluorescence. An analogous method utilizing light emission by luciferase was recently published (23). Both fluorescence resonance energy transfer (FRET) (24-26) and bioluminescence resonance energy transfer (BRET) (27) are based on donor molecules exciting acceptor molecules that then emit light. Only when the donor and acceptor are in close proximity, within a few nm and in a favorable orientation, will resonance energy transfer occur.

Although methods such as FRET, BRET and BiFC are very efficient and widely used for interaction studies in living cells, the non-physiological levels of expression of the transgenes, along with the risk that properties of the fusion-proteins may differ from those of the native proteins, may seriously influence the results. However, until now, no methods have been available for the detection of endogenous protein interactions *in situ*.

DETECTION OF PROTEIN INTERACTIONS AND MODIFICATIONS BY PROXIMITY LIGATION

The properties of proximity ligation make it ideal also to detect and measure protein interactions and modifications. By using two or more antibodies directed against interacting partner proteins, the interacting molecules can be detected in a homogenous assay allowing detection of low abundant molecules or rare interactions (Gustafsdottir et al., in progress). For *in situ* detection, the P-LISA method results in highly-specific and strongly-amplified signals, allowing detection of individual endogenous protein interactions (15) or detection of post-translationally modified proteins, such as phosphorylated receptors with little or no background (Jarvius et al., unpublished). An additional benefit is that the technique enables multiplexed detection, as different proximity probes can give rise to distinct RCA products detectable using specific fluorescence labeled oligonucleotide probes.

SUMMARY

The purpose of this review has been a discussion of different methods for detection of proteins and protein-protein interactions from the point of view of specificity of analysis, focusing on proximity ligation. As all methods have their pros and cons, the choice of method must depend on the question that needs to be addressed, taking in account the time required for the analysis, cost of reagents and availability of instruments, etc. Table 1 summarizes the different methods discussed in this review.

The methods can be divided into three groups according to whether the proteins are analyzed using antibodies (e.g., ELISA, Western blot, immunohistochemistry, flow cytometry, co-immunoprecipitation and proximity ligation), by constructing ectopically expressed fluorescent proteins (FRET/BRET and BiFC), or using mass spectrometry. Future development of methods will depend on further biotechnological progress, but also upon availability of antibodies against all proteins expressed in humans (28) (http://www.proteinatlas.org). The recent development of mass spectrometry and imaging mass spectrometry (reviewed by Chaurand et al. (29) provide us with a tool to investigate protein expression within tissues in a hypothesis-free manner, allowing detection also of proteins to which there are no antibodies. However, improvements in resolution, currently in the 50 μm range, and increased sensitivity will be necessary to allow detection of low abundant proteins.

The completion of the human genome project and the emergence of new molecular tools to study biomolecules and their interactions will fundamentally impact our understanding of life and pathology. With the human genome

Table 1. Utility of different techniques for protein detection.

	Protein identification	Localized detection	Endogenous proteins	Protein interactions	Analyzing living cells
ELISA/protein array	+	–	+	+	–
Western blot	+	–	+	–	–
Mass spectrometry	+	+/–	+	–	–
Immunohistochemistry	+/–	+	+	–	–
co-immunoprecipitation	+	–	+	+	–
FRET/BRET/BiFC	–	+	–	+	+
Proximity ligation	+	+	+	+	–

mapped at maximal resolution, the task for the future is now to understand the exceedingly complex interplay between all gene products. Returning to the metaphor of the birds, a cell corresponds to a very richly populated lake indeed, inhabited by many different species busily interacting in different combinations. We can now anticipate having the binoculars to spot a swan in the midst of flocks of all other birds, and to be in a position to observe its day-to-day interactions with its partners and any passers-by.

ACKNOWLEDGMENTS

The work in our laboratory is supported by grants from the Wallenberg Foundation, by the EU Integrated Project "MolTools" and the Research Councils of Sweden for natural science and medicine.

REFERENCES

1 Wide, L., Bennich, H. and Johansson, S.G. (1967) Lancet 2(7526), 1105-1107.
2 Engvall, E. and Perlman, P. (1971) Immunochemistry 8(9), 871-874.
3 Coons. A.H., Creech, H.J., Jones, R.N. and Berliner, E. (1942) J. Immunol. 45, 159-170.
4 Nakane, P.K. and Pierce, G.B., Jr. (1967) J. Cell Biol. 33(2), 307-318.
5 Sano, T.S.C. and Cantor, C.R. (1992) Science 258(5079), 120-122.
6 Schweitzer, B., Wiltshire, S., Lambert, J., O'Malley, S., Kukanskis, K., Zhu, Z., Kingsmore, S.F., Lizardi, P.M. and Ward, D.C. (2000) Proc. Nat. Acad. Sci. U.S.A. 97(18), 10113-10119.
7 Wiltshire, S., O'Malley, S., Lambert, J., Kukanskis, K., Edgar, D., Kingsmore, S.F. and Schweitzer, B. (2000) Clin. Chem. 46(12), 1990-1993.
8 Gusev, Y., Sparkowski, J., Raghunathan, A., Ferguson, H., Jr., Montano, J., Bogdan, N., Schweitzer, B., Wiltshire, S., Kingsmore, S.F., Maltzman, W. and Wheeler, V. (2001) Amer. J. Pathol. 159(1), 63-69.
9 Espina, V., Woodhouse, E.C., Wulfkuhle, J., Asmussen, H.D., Petricoin, E.F., 3rd and Liotta, L.A. (2004) J. Immunol. Methods 290(1-2), 121-133.
10 Brody, E.N., Willis, M.C., Smith, J.D., Jayasena, S., Zichi, D. and Gold, L. (1999) Mol. Diagn. 4(4), 381-388.
11 Fredriksson, S., Gullberg, M., Jarvius, J., Olsson, C., Pietras, K., Gustafsdottir, S.M., Östman, A. and Landegren, U. (2002) Nat. Biotechnol. 20(5), 473-477.
12 Gullberg, M., Gustafsdottir, S.M., Schallmeiner, E., Jarvius, J., Bjarnegard, M., Betsholtz, C., Landegren, U. and Fredriksson, S. (2004) Proc. Nat. Acad. Sci. U.S.A. 101(22), 8420-8424.
13 Gustafsdottir, S.M., Nordengrahn, A., Fredriksson, S., Wallgren, P., Rivera, E., Schallmeiner, E., Merza, M. and Landegren, U. (2006) Clin. Chem. 52(6):1152-1160.
14 Larsson, C., Koch, J.E., Nygren, A., Janssen, G., Raap, A.K., Landegren, U. and Nilsson, M. (2004) Nat. Methods 1(3), 227-232.

15 Söderberg, O., Gullberg, M., Jarvius, M., Ridderstråle, K., Leuchowius, K-J., Jarvius, J., Wester, K., Hydbring, P., Bahram, F., Larsson, L-G. and Landegren, U. Nature Methods, *in press.*

16 Hell, S.W. (2003) Nat. Biotechnol. 21(11), 1347-1355.

17 Fields, S. and Song, O. (1989) Nature 340(6230), 245-246.

18 Chien, C.T., Bartel, P.L., Sternglanz, R. and Fields, S. (1991) Proc. Nat. Acad. Sci. U.S.A. 88(21), 9578-9582.

19 Rossi, F., Charlton, C.A. and Blau, H.M. (1997) Proc. Nat. Acad. Sci. U.S.A. 94(16), 8405-8410.

20 Pelletier, J.N., Campbell-Valois, F.X. and Michnick, S.W. (1998) Proc. Nat. Acad. Sci. U.S.A. 95(21), 12141-12146.

21 Johnsson, N. and Varshavsky, A. (1994) Proc. Nat. Acad. Sci. U.S.A. 91(22), 10340-10344.

22 Hu, C.D., Chinenov, Y. and Kerppola, T.K. (2002) Mol. Cell 9(4), 789-798.

23 Paulmurugan, R. and Gambhir, S.S. (2003) Anal. Chem. 75(7), 1584-1589.

24 Mitra, R.D., Silva, C.M. and Youvan, D.C. (1996) Gene 173(1 Spec No), 13-17.

25 Mahajan, N.P., Linder, K., Berry, G., Gordon, G.W., Heim, R. and Herman, B. (1998) Nat. Biotechnol. 16(6), 547-552.

26 Galperin, E., Verkhusha, V.V. and Sorkin, A. (2004) Nat. Methods 1(3), 209-217 Epub 2004 Nov 2018.

27 Xu, Y., Piston, D.W. and Johnson, C.H. (1999) Proc. Nat. Acad. Sci. U.S.A. 96(1), 151-156.

28 Uhlén, M., Björling, E., Agaton, C., Szigyarto, C.A., Amini, B., Andersen, E., Andersson, A.C., Angelidou, P., Asplund, A., Asplund, C., Berglund, L., Bergström, K., Brumer, H., Cerjan, D., Ekström, M., Elobeid, A., Eriksson, C., Fagerberg, L., Falk, R., Fall, J., Forsberg, M., Björklund, M.G., Gumbel, K., Halimi, A., Hallin, I., Hamsten, C., Hansson, M., Hedhammar, M., Hercules, G. and Kampf, C. (2005) Mol. Cell Proteomics Aug 27; [Epub ahead of print].

29 Chaurand, P., Sanders, M.E., Jensen, R.A. and Caprioli, R.M. (2004) Amer. J. Pathol. 165(4), 1057-1068.

PROTEIN OVEREXPRESSION IN MAMMALIAN CELL LINES

Paul Freimuth

Biology Department
Brookhaven National Laboratory
Upton, NY 11973

INTRODUCTION

Large scale production of mammalian proteins is required for many diagnostic and therapeutic applications and also for structural characterization and other basic studies. Soluble mammalian proteins often can be overexpressed in bacterial cells, but in many cases these proteins misfold or do not exhibit proper function due to lack of necessary posttranslational modifications. Perhaps the most severe difficulties arise in the expression of membrane proteins, where almost invariably mammalian proteins fail to insert in proper functional form in the membranes of bacteria or other heterologous host cells. For example, less than 100 membrane protein structures have been determined and of these only about 5 were derived from eukaryotic sources (1). By contrast structures have been determined for several thousand soluble proteins. Two major technical obstacles to solving membrane protein structures are production of sufficient amounts of protein for analysis and stabilization of protein conformation following extraction from the lipid bilayer. Stabilization is a problem common

Genetic Engineering, Volume 28, Edited by J. K. Setlow

to membrane proteins from both prokaryotic and eukaryotic cells; thus the relatively few structures of eukaryotic membrane proteins can largely be attributed to difficulties in producing these proteins in sufficient quantity for crystallization.

Recent development of methods for crystal growth in nanoliter-scale droplets (2-4) and the availability of ultrabright-synchrotron-radiation-sources for diffraction analysis of very small crystals (5) should reduce the absolute quantities of protein needed for X-ray crystallography. Despite these advances, structure determination still requires access to milligram amounts of highly-purified-protein. Bacterial expression systems are capable of meeting these demands, as attested to by the rapid increase in the number of prokaryotic membrane protein structures that have been solved. Unfortunately, the difficulties in expression of eukaryotic membrane proteins in bacteria leave the options of isolating these proteins either from a natural tissue source or from cultured eukaryotic cells. One additional potential solution to this problem is to express proteins in transgenic animals, with the current aim typically of directing the protein to accumulate in milk, egg white or blood. The use of transgenic animals for this purpose has been reviewed recently (6) and will not be discussed further here except to note that many of the difficulties with stable protein expression in cell lines, discussed further below, also can occur in transgenic animals.

TRANSIENT *VS* STABLE EXPRESSION

One question to consider at the outset is whether it is reasonable to expect that milligram quantities of a membrane protein, or also of soluble proteins for that matter, could be produced in cultured mammalian cells using equipment generally available in research-scale-laboratories. If we consider a hypothetical case where cultured cells can be engineered to express 10^6 copies per cell of a 50 kD membrane protein, then about 10^{10} cells would yield 1 mg of protein. Growth of mammalian cell lines at this scale, although not routine, is within the capabilities of many research labs. Furthermore, suspension-adapted cell lines such as HeLa S3 and CHO (Chinese Hamster Ovary) can be grown at much larger scale using specialized fermentation equipment that is available to the research community at cell-culture-resource-facilities such as those operated by the NIH. The goal of producing milligram quantities of membrane proteins and also soluble proteins in mammalian cells therefore could be achieved using existing cell lines and culture technologies.

A current limitation of this approach is the difficulty in engineering stable gene expression in mammalian cell lines. These difficulties potentially can be circumvented by transient expression immediately following introduction of the genes of interest into cells by one of several transfection procedures or by transduction with viral vectors. Transient expression occurs only within cells that initially take up the exogenous DNA, and cells are generally harvested within a few days after transfection for isolation of the expressed protein. Although the frequency of productively-transfected cells has been increased substantially by recent development of highly effective DNA transfection agents such as cationic liposomes and others (7), it would be difficult to scale up these protocols to transfect the very large number of cells that would be needed to express milligram quantities of protein. On the other hand, transduction of large-scale-cell-cultures with either vaccinia virus (8) or adenovirus (9) based vectors is possible since both viruses can be readily produced in very large quantities. Transduction with viral vectors is beyond the scope of this article but may be a viable approach for research laboratories that are equipped for large-scale infections.

Stable integration of cDNA or gene constructs into the genome of a host cell has several potential advantages over transient expression. In principle, after initial derivation of cell lines that overexpress the target proteins of interest, the cells can be grown continuously until the required number of cells has been accumulated. Approximately 30 generations are necessary to produce 10^{10} cells, about one month of continuous culture for most cell lines. In practice, however, initial levels of protein expression often decay to background levels after just a few cell generations, resulting from transcriptional silencing of the integrated exogenous DNA (10). Silencing can occur by different mechanisms, including DNA hypermethylation, histone deacetylation, and formation of condensed chromatin across the locus of integration. These mechanisms are used by mammalian cells not only to regulate endogenous genes but also to block expression of parasitic elements such as retroviral provirus genomes and retrotransposons that have accumulated in the genome in large numbers during evolution (11). Clearly cells must have the means to discriminate parasitic sequences, including exogenous transfected genes, from endogenous housekeeping genes that must be maintained in a transcriptionally active, open chromatin conformation to insure continued expression and cell survival.

Although current understanding of gene regulation in the context of mammalian cell chromatin is far from complete, investigations of several model genes have led to the discovery of regulatory elements that can block silencing mechanisms when the genes are introduced into transgenic animals or stable cell lines (12). These elements are of two types, transcription enhancers and boundary/insulator elements. The human β-globin locus consists of several genes spread across a ~100 kb region that are expressed at different stages of development and are regulated by a single large enhancer known as the locus control region (LCR) (13,14). The LCR itself spans about 30 kb and is located upstream of the globin gene cluster. In certain β-thalassemias, the globin genes are intact but not expressed because of deletions that extend into the LCR region. The globin genes in these mutant cells reside in condensed chromatin, as opposed to the open chromatin (DNase-sensitive) conformation of the β-globin locus in normal cells (15,16). Consistent with this observation, early studies showed that a 5 kb fragment containing the β-globin gene and its promoter was not expressed or expressed at lower than normal levels in transgenic mice, but that addition of LCR elements enabled this gene to be expressed at normal levels and in an appropriate tissue-specific pattern (17). Importantly, the LCR elements also enabled expression of the β-globin gene in a copy number-dependent and position-independent manner in transgenic mice.

Further investigations led to the discovery that the open, DNase-sensitive conformation of the β-globin locus does not result from an activity of the LCR itself but from the activity of other elements termed boundary or insulator elements that flank the globin locus (18,19). Elements with boundary/insulator function recently have come under intense scrutiny and have been characterized in detail in *Drosophila* (gypsy element), in association with the chick β-globin locus and in association with the mouse and human c-myc genes, to name a few (20). These elements define the boundary between condensed and open chromatin as defined by the presence of methylated *vs* acetylated histones, respectively. Insulators also can block enhancer function when interposed between a promoter and an enhancer. Enhancer-blocking function was shown to require the binding of a transcription factor, CTCF, to the insulator elements in both the c-myc and β-globin genes (21,22). CTCF-binding insulators have now been identified associated with over 200 known genes, using chromatin immunoprecipitation assays and whole genome microarray analysis (23). These analyses have led to the derivation of a consensus DNA sequence for CTCF binding (21).

A model for active chromatin domains begins to emerge from these studies (24), where transcribed genes are localized within loops of chromatin that are anchored to the nuclear matrix or nucleolar surface at either end by matrix association elements. Boundary/insulator elements are located within the loop, flanking the gene and its promoter/enhancer elements. The c-myc gene appears to have a single boundary/insulator element located upstream of the promoter region (24). Insulator elements generally are not located between the promoter and enhancer elements within a single transcription unit. At this point it is not clear how the β-globin LCR confers copy number-dependent and position-independent expression on linked transgenes in transgenic mice, since the LCR has not been shown to have boundary element function and thus should not be able to block formation of silencing heterochromatin at loci of transgene integration. Clearly the regulation of gene expression in the context of chromatin is complex, and the action of elements when inserted at ectopic sites understandably may be difficult to predict.

As elements that regulate expression of model endogenous genes such as the β-globin locus have been identified and characterized, their effects on the expression of exogenous cDNAs in transfected or virally transduced cells and in transgenic animals have been evaluated. Excellent reviews that describe these studies in detail both in the context of transgenic animals (25) and cell lines (26) have been published recently. Most regulatory elements that have been studied to date, and those reviewed above, were derived from genes that have tissue-specific expression patterns; however, elements with similar properties have not been found in association with ubiquitously expressed housekeeping genes. In their investigation of the regulation of housekeeping genes, Antoniou and colleagues found that short genomic fragments containing the promoter and CpG island region had chromatin-opening function, and that this activity was dominant if the genomic fragment contained a CpG island spanning dual, divergently-transcribed promoters (27). Reporter transgenes cloned behind these promoter fragments exhibited stable long-term expression even when integrated in centromeric heterochromatin. The authors suggest that open chromatin may be established during transcription elongation of these genes, and that divergent transcription thus might decrease transcriptional interference from a weak single promoter.

EXPRESSION FROM CLONED GENOMIC FRAGMENTS

Construction of physical maps of chromosomes was an important aspect of the human genome project, and the genomic fragments used for mapping often were cloned in bacterial artificial chromosome (BAC) vectors (28). The average size of genomic DNA inserts in these BAC libraries was about 150 kb, consequently many human genes of <50 kb will be isolated on single BAC clones. BAC clones containing intact human genes offer potential advantages over cDNA expression vectors that are typically used for transgenic protein expression in mammalian cells. First, the gene will be transcribed from the endogenous promoter and also may be regulated by endogenous elements contained within the gene or its flanking region. The large size of these cloned fragments also might provide fortuitous boundary function, shielding the gene or promoter region from encroachment of silencing heterochromatin. By contrast, the large size of these clones limits the ability to manipulate the gene sequence or to introduce reporters or selectable markers into the clone. Several methods to retrofit BACs with reporter genes, selectable markers or replication origins using recombination based strategies have recently been described (29-32). These recombination strategies also can be used to construct specific mutations in coding or regulatory sequences of the cloned gene.

Difficulties in preparation and handling of large BAC clones also can result in low transfection efficiencies. Since BACs are derived from F factor, they replicate to only 1 or 2 copies per cell. Large amounts of culture must be processed to purify BAC DNA compared to amounts needed for plasmid purification, and the BAC preparations therefore may be less pure than typical plasmid preps. Impurities in BAC DNA preps can interfere with formation of proper-sized particles for transfection, leading to low transformation efficiencies (33). These problems can be avoided by using more extensive purification procedures for BAC DNA. An alternative approach was recently described where intact *E. coli* cells containing BAC DNA were taken up by mammalian cells through receptor-mediated endocytosis (30). The *E. coli* strain used for this procedure expressed the *Yersinia* invasin protein, which interacts with mammalian cell integrin receptors, and also had a mutation that weakened the bacterial cell wall, thus allowing more efficient release of BAC DNA upon entry into the mammalian cell.

A recent study described co-transfection of CHO cells with a mixture of 2 BAC clones, one containing the selectable mouse *Dhfr* gene and the other containing the intact mouse Cdc6 gene (34). DNAs were transfected into cells as Lipofectamine complexes. This study led to several notable results: co-transformation frequencies of about 80% were achieved; cell clones with stable integration of up to 15 copies of the Cdc6 BAC were isolated; levels of mouse Cdc6 protein produced were proportional to the BAC copy number integrated in the cell clones; and expression levels of Cdc6 protein were constant for at least one year of continuous cell culture. The key finding of this study, however, was that the mouse Cdc6 gene was regulated in an appropriate cell cycle-depended manner in the transfected clones. This particular case thus serves as perhaps a best case scenario of all the potential advantages that can result from transfection of intact genes with their endogenous promoters and regulatory signals as opposed to transfection of cDNA expression constructs with heterologous promoters and signals.

EXPRESSION OF THE HUMAN *CXADR* GENE IN BAC-TRANSFECTED MOUSE CELLS

Studies in our lab have focused on expression of the human coxsackie and adenovirus receptor (CAR) protein in mouse cells transfected with a BAC clone of the intact CAR gene (*CXADR*). Subgroup B coxsackieviruses and the majority of adenovirus serotypes bind to CAR (35), a 46 kD glycoprotein member of the immunoglobulin (IG) superfamily (36, 37) that is highly conserved in vertebrates (38). CAR is expressed in many human and rodent tissues (36, 39-41), and thus can be considered a ubiquitous housekeeping gene. CAR has two extracellular IgG-type domains, a single membrane-spanning region, and a 110 amino acid cytoplasmic tail. The cytoplasmic tail contains several motifs that direct CAR to the basolateral surfaces of polarized epithelial cells (42,43), where it localizes specifically in tight junctions (44, 45). We earlier mapped the *CXADR* gene to a single locus on human chromosome 21 (46) and isolated several BAC clones that hybridized to probes derived from this locus. To screen for clones that contained a functional *CXADR* gene, mouse A9 cells, which do not express CAR protein, were transfected with calcium phosphate co-precipitates of BAC DNA and plasmid DNA containing the selectable neomycin phosphotransferase gene. In transfections with one BAC clone, about 50% of G418-resistant colonies expressed CAR protein on the cell surface as determined with a CAR-specific erythrocyte rosette assay (46). The 120 kb genomic DNA insert of this clone was sequenced (GenBank accession AF200465) and found to contain the complete 60 kb *CXADR* gene plus about 30 kb of upstream and downstream flanking sequence.

Several CAR-expressing BAC-transformed A9 cell clones were expanded in culture and characterized for BAC copy number and levels of CAR protein and mRNA. The concentration of CAR protein expressed on the surface of these clones varied over a wide range, and in most cases exceeded the levels of CAR expressed on the surface of HeLa cells (Figure 1A). Importantly, CAR protein levels were roughly proportional to the copy number of BAC DNA integrated into the cell genome and to the abundance of CAR mRNA (Figure 1B). CAR protein levels were constant over several months of continuous culture for some of these clones, and one clone (#5) was used as the source of human CAR protein for structural characterization of coxsackievirus-CAR interaction by electron cryomicroscopy (47).

The stable, copy-number-dependent expression of CAR observed in these clones suggests that the *CXADR* gene or flanking regions contain elements that block gene silencing. We have not conducted any further studies to isolate and characterize such elements, but considering that *CXADR* is a ubiquitous housekeeping gene, then the CpG island and promoter region alone might be sufficient to block silencing. However, the completed human genome sequence shows that there is no confirmed expressed gene within 1 Mb upstream of the *CXADR* gene, and that the 3′ ends of the convergently transcribed *CXADR* and *ANA* genes are

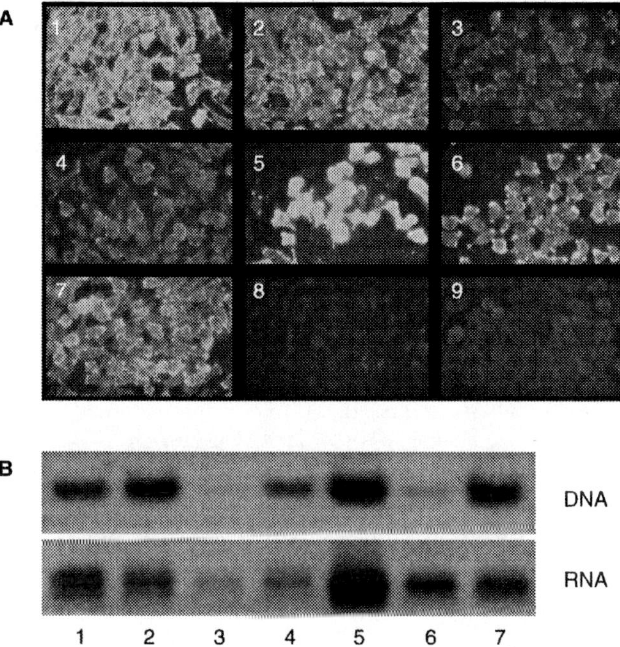

Figure 1. Expression of human *CXADR* in BAC-transfected mouse A9 cells. (A), fluorescence microscopy of 7 BAC-transfected mouse A9 cell clones (panels 1-7), parental mouse A9 cells (panel 8) and human HeLa cells (panel 9) after sequential incubation with biotin-labeled fiber knob domain from adenovirus-12 (50) and anti-biotin-fluorescein conjugate. (B), Southern blot (upper panel labeled DNA) and Northern blot (lower panel labeled RNA) of equal amounts of genomic DNA or total cytoplasmic RNA, respectively, extracted from clones 1-7 (lanes 1-7) probed with ^{32}P-labeled human CAR cDNA. Reproduced from Figure 5 on page 348 of (51) with kind permission of Spring Science and Business Media.

separated by about 23 kb. Thus it is likely that chromatin on at least the upstream side of the *CXADR* gene is condensed, possibly necessitating the presence of an insulator or boundary element to block encroachment of silencing heterochromatin into the *CXADR* gene. Interestingly, sequences matching the CTCF consensus binding site associated with insulator elements are located just upstream of the CpG island containing the *CXADR* promoter and also within the *CXADR-ANA* intergenic region.

Similar to the results reported by Heintz and colleagues (33), we found that transfection efficiencies with the CXADR BAC clone depended critically on the purity of the BAC DNA. In our case, DNAs were introduced into cells as calcium phosphate precipitates, whereas the experiments of Heintz used cationic liposomes (34). We isolated BAC DNAs from 250 ml cultures of *E. coli* by alkaline lysis (48) using solutions I, II and III as described (49). BAC DNAs were purified by isopropanol precipitation, RNase digestion, sequential extraction with phenol, phenol:chloroform and chloroform, and final precipitation with ethanol according to a protocol formerly posted on the Whitehead Institute Genome Center web site. Calcium phosphate precipitates of BAC DNA prepared by this method did not form large aggregates and transfection efficiencies were comparable to control transfections where an equivalent amount of commercial salmon sperm DNA was substituted for the BAC DNA. Remarkably, in both our experiment and that reported by Heintz, cell clones were isolated that had stably integrated more than 10 copies of the intact, functional gene—probably in excess of 2 Mb of exogenous DNA. Rodent cell lines were used as transfection recipients in both studies, and these may have a greater capacity than human cell lines to maintain large amounts of integrated exogenous DNA.

CONCLUSIONS

Basic studies of the regulation of gene expression in the context of chromatin have led to significant improvements in methods for stable protein expression in mammalian cell lines. If this trend continues, then protein expression may no longer be an obstacle to structural characterization, particularly of membrane proteins. BAC transfection appears to be an attractive and underused approach for dissection and functional characterization of endogenous regulatory elements, and also may prove to be the most robust and direct approach for stable overexpression of proteins in transgenic animals and in transfected mammalian cells.

ACKNOWLEDGMENTS

Studies in our laboratory on CAR expression were supported by a grant from the USPHS.

REFERENCES

1 White, S.H. (2004) Protein Sci. 13, 1948-1949.
2 Hansen, C.L., Skordalakes, E., Berger, J.M. and Quake, S.R. (2002) Proc. Nat. Acad. Sci. U.S.A. 99, 16531-16536.
3 Kuil, M.E., Bodenstaff, E.R., Hoedemaeker, F.J., Abrahams, J.P., DeLucas, L.J., Bray, T.L., Nagy, L., McCombs, D., Chernov, N., Hamrick, D., Cosenza, L., Belgovskiy, A., Stoops, B., Chait, A., Hansen, C.L., Skordalakes, E., Berger, J.M. and Quake, S.R. (2002) Enzyme Microb. Technol. 30, 262-265.
4 DeLucas, L.J., Bray, T.L., Nagy, L., McCombs, D., Chernov, N., Hamrick, D., Cosenza, L., Belgovskiy, A., Stoops, B. and Chait, A. (2003) J. Struct. Biol. 142, 188-206.
5 Dauter, Z. (2005) Prog. Biophys. Mol. Biol. 89, 153-172.
6 Houdebine, L.M. (2000) Transgenic Res. 9, 305-320.
7 Thomas, P. and Smart, T.G. (2005) J. Pharmacol. Toxicol. Methods 51, 187-200.
8 Mackett, M., Smith, G.L. and Moss, B. (1992) Biotechnology 24, 495-499.
9 Hitt, M.M., Addison, C.L. and Graham, F.L. (1997) Adv. Pharmacol. 40, 137-206.

10 Kwaks, T.H., Barnett, P., Hemrika, W., Siersma, T., Sewalt, R.G., Satijn, D.P., Brons, J.F., van Blokland, R., Kwakman, P., Kruckeberg, A.L., Kelder, A. and Otte, A.P. (2003) Nat. Biotechnol. 21, 553-558.

11 Lorincz, M.C., Schubeler, D. and Groudine, M. (2001) Mol. Cell Biol. 21, 7913-7922.

12 Kioussis, D. and Festenstein, R. (1997) Curr. Opin. Genet. Dev. 7, 614-619.

13 Bulger, M. and Groudine, M. (1999) Genes Dev. 13, 2465-2477.

14 Li, Q., Peterson, K.R., Fang, X. and Stamatoyannopoulos, G. (2002) Blood 100, 3077-3086.

15 Forrester, W.C., Takegawa, S., Papayannopoulou, T., Stamatoyannopoulos, G. and Groudine, M. (1987) Nucl. Acids Res. 15, 10159-10177.

16 Forrester, W.C., Epner, E., Driscoll, M.C., Enver, T., Brice, M., Papayannopoulou, T. and Groudine, M. (1990) Genes Dev. 4, 1637-1649.

17 Grosveld, F., van Assendelft, G.B., Greaves, D.R. and Kollias, G. (1987) Cell 51, 975-985.

18 Reik, A., Telling, A., Zitnik, G., Cimbora, D., Epner, E. and Groudine, M. (1998) Mol. Cell Biol. 18, 5992-6000.

19 Epner, E., Reik, A., Cimbora, D., Telling, A., Bender, M.A., Fiering, S., Enver, T., Martin, D.I., Kennedy, M., Keller, G. and Groudine, M. (1998) Mol. Cell 2, 447-455.

20 West, A.G., Gaszner, M. and Felsenfeld, G. (2002) Genes Dev. 16, 271-288.

21 Bell, A.C., West, A.G. and Felsenfeld, G. (1999) Cell 98, 387-396.

22 Filippova, G.N., Fagerlie, S., Klenova, E.M., Myers, C., Dehner, Y., Goodwin, G., Neiman, P.E., Collins, S.J. and Lobanenkov, V.V. (1996) Mol. Cell Biol. 16, 2802-2813.

23 Mukhopadhyay, R., Yu, W., Whitehead, J., Xu, J., Lezcano, M., Pack, S., Kanduri, C., Kanduri, M., Ginjala, V., Vostrov, A., Quitschke, W., Chernukhin, I., Klenova, E., Lobanenkov, V. and Ohlsson, R. (2004) Genome Res. 14, 1594-1602.

24 Gombert, W.M., Farris, S.D., Rubio, E.D., Morey-Rosler, K.M., Schubach, W.H. and Krumm, A. (2003) Mol. Cell Biol. 23, 9338-9348.

25 Recillas-Targa, F., Valadez-Graham, V. and Farrell, C.M. (2004) Bioessays 26, 796-807.

26 Barnes, L.M., Bentley, C.M. and Dickson, A.J. (2003) Biotechnol. Bioeng. 81, 631-639.

27 Antoniou, M., Harland, L., Mustoe, T., Williams, S., Holdstock, J., Yague, E., Mulcahy, T., Griffiths, M., Edwards, S., Ioannou, P.A., Mountain, A. and Crombie, R. (2003) Genomics 82, 269-279.

28 Shizuya, H., Birren, B., Kim, U.-J., Mancino, V., Slepak, T., Tachiiri, Y. and Simon, M. (1992) Proc. Nat. Acad. Sci. U.S.A. 89, 8794-8797.

29 Orford, M., Nefedov, M., Vadolas, J., Zaibak, F., Williamson, R. and Ioannou, P.A. (2000) Nucl. Acids Res. 28, e84.

30 Narayanan, K. and Warburton, P.E. (2003) Nucl. Acids Res. 31, e51.

31 Magin-Lachmann, C., Kotzamanis, G., D'Aiuto, L., Wagner, E. and Huxley, C. (2003) BMC Biotechnol. 3, 2.

32 Wang, Z., Engler, P., Longacre, A. and Storb, U. (2001) Genome Res. 11, 137-142.

33 Montigny, W.J., Phelps, S.F., Illenye, S. and Heintz, N.H. (2003) Biotechniques 35, 796-807.

34 Illenye, S. and Heintz, N.H. (2004) Genomics 83, 66-75.

35 Roelvink, P.W., Lizonova, A., Lee, J.G., Li, Y., Bergelson, J.M., Finberg, R.W., Brough, D.E., Kovesdi, I. and Wickham, T.J. (1998) J. Virol. 72, 7909-7915.

36 Tomko, R.P., Xu, R. and Philipson, L. (1997) Proc. Nat. Acad. Sci. U.S.A. 94, 3352-3356.

37 Bergelson, J.M., Cunningham, J.A., Droguett, G., Kurt-Jones, E.A., Krithivas, A., Hong, J.S., Horwitz, M.S., Crowell, R.L. and Finberg, R.W. (1997) Science 275, 1320-1323.

38 Chretien, I., Marcuz, A., Courtet, M., Katevuo, K., Vainio, O., Heath, J.K., White, S.J. and Du Pasquier, L. (1998) Eur. J. Immunol. 28, 4094-4104.

39 Fechner, H., Haack, A., Wang, H., Wang, X., Eizema, K., Pauschinger, M., Schoemaker, R., Veghel, R., Houtsmuller, A., Schultheiss, H.P., Lamers, J. and Poller, W. (1999) Gene Ther. 6, 1520-1535.

40 Tomko, R.P., Johansson, C.B., Totrov, M., Abagyan, R., Frisen, J. and Philipson, L. (2000) Exp. Cell Res. 255, 47-55.

41 Bergelson, J.M., Krithivas, A., Celi, L., Droguett, G., Horwitz, M.S., Wickham, T., Crowell, R.L. and Finberg, R.W. (1998) J. Virol. 72, 415-419.

42 Walters, R.W., Grunst, T., Bergelson, J.M., Finberg, R.W., Welsh, M.J. and Zabner, J. (1999) J. Biol. Chem. 274, 10219-10226.

43 Cohen, C.J., Gaetz, J., Ohman, T. and Bergelson, J.M. (2001) J. Biol. Chem. 276, 25392-25398.

44 Walters, R.W., Freimuth, P., Moninger, T.O., Ganske, I., Zabner, J. and Welsh, M.J. (2002) Cell 110, 789-799.

45 Cohen, C.J., Shieh, J.T., Pickles, R.J., Okegawa, T., Hsieh, J.T. and Bergelson, J.M. (2001) Proc. Nat. Acad. Sci. U.S.A. 98, 15191-15196.

46 Mayr, G.A. and Freimuth, P. (1997) J. Virol. 71, 412-418.

47 He, Y., Chipman, P.R., Howitt, J., Bator, C.M., Whitt, M.A., Baker, T.S., Kuhn, R.J., Anderson, C.W., Freimuth, P. and Rossmann, M.G. (2001) Nat. Struct. Biol. 8, 874-878.

48 Birnboim, H.C. and Doly, J. (1979) Nucl. Acids Res. 7, 1513-1523.

49 Sambrook, J., Fritsch, E. and Maniatis, T. (1989) Molecular Cloning: A Laboratory Manual, Cold Spring Harbor Laboratory Press, Cold Spring Harbor, NY.

50 Freimuth, P., Springer, K., Berard, C., Hainfeld, J., Bewley, M. and Flanagan, J. (1999) J. Virol. 73, 1392-1398.

51 Howitt, J., Anderson, C.W. and Freimuth, P. (2003) Adenovirus interaction with its cellular receptor CAR Curr. Top. Microbiol. Immunol. 272, 331-364.

A HIGH-THROUGHPUT APPROACH TO PROTEIN STRUCTURE ANALYSIS

Babu A. Manjasetty[1,2], Wuxian Shi[1,2], Chenyang Zhan[3], András Fiser[3] and Mark R. Chance[1,2,3*]

[1]New York Structural GenomiX Research Consortium (NYSGXRC)
Center for Synchrotron Biosciences
National Synchrotron Light Source
Brookhaven National Laboratory
Upton NY 11973
[2]Case Center for Proteomics
Case Western Reserve University
10900, Euclid Avenue
Cleveland, OH 44106
[3]New York Structural GenomiX Research Consortium (NYSGXRC)
Department of Biochemistry
Albert Einstein College of Medicine
Bronx, NY 10461

*To whom correspondence should be addressed.

Genetic Engineering, Volume 28, Edited by J. K. Setlow

INTRODUCTION

Structural Genomics was defined at the 2nd International Structural Genomics conference in 2001 as, "A large-scale project to determine the three-dimensional shapes of all proteins and other important biological molecules encoded by the genomes of key organisms". The structural genomics projects aim at the discovery, analysis and dissemination of three-dimensional structures of all proteins and other biological macromolecules in the universe of protein folds (Figure 1). The major structural genomics initiatives around the world are listed in Table 1.

The Protein Structure Initiative (PSI), which comprise the major efforts in structural genomics in the United States, has established centers for the project that have achieved automation of all the steps involved in determining protein structures, including target selection, cloning, expression, purification, biophysical characterization, crystallization, data collection, structure solution, refinement, validation and functional annotation (Figure 2). In addition, international coordination was put in place among the different centers in the U.S.A. and worldwide to avoid duplication of efforts and waste of resources (http://targetdb.pdb.org/).

Among the many structural genomics research projects around the world, in the U.S.A., the National Institutes of General Medical Science (NIGMS) of the National Institutes of Health (NIH) sponsored nine pilot structural genomics centers through the first phase of PSI (1). During this pilot phase (PSI1), these centers have established infrastructure for high-throughput production of protein

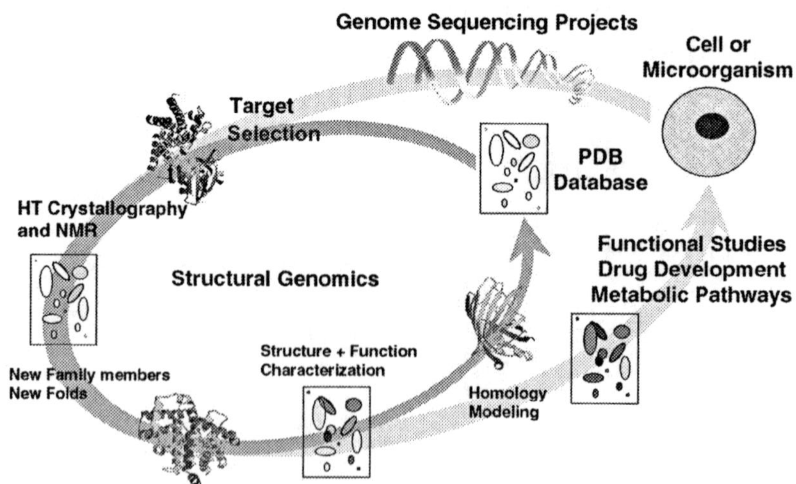

Figure 1. An example representing global structural genomics efforts for completing the protein family and fold landscape. The rectangular panels represent our current knowledge of the set of protein sequence families, showing whether they contain any 3D structural examples (black encircled regions) or not (white encircled regions). The amount of black increases as more structures are determined experimentally. Only a small fraction of the protein families may not contain a known 3D structure (small circles), but the majority of the fold landscape will be represented, permitting homology modeling of most of the remaining and new gene sequences. Diagram taken from Stevens, R.C., Yokoyama, S., Wilson, I.A. (2001) Global efforts in structural genomics, Science, 294, 89-92.

Table 1. Major structural Genomics Centers around the world.

Country	Initiative	Web Address
Japan	RIKEN Structural Genomics Initiative (RSGI)	http://www.riken.go.jp/engn/index.html
England	Structural Proteomics in Europe (SPINE)	http://www.spineurope.org/page.php?page=home
	Oxford Protein Production Facility (OPPF)	http://www.oppf.ox.ac.uk/index.php?module=ContentExpress&func=display&ceid=1&meid=-1
U.S.A.	NIGMS Protein Structure Initiative (PSI)	http://www.nigms.nih.gov/Initiatives/PSI
Canada	Montreal-Kingston Bacterial Genomics Initiative (BSGI)	http://euler.bri.nrc.ca/brimsg/bsgi.html
Germany	Protein Structure Factory (PSF)	http://www.proteinstrukturfabrik.de/
	Mycobacterium Tuberculosis Structural Proteomics Project (XMTB)	http://xmtb.org/start.html
Israel	The Israel Structural Proteomics Center	http://www.weizmann.ac.il/ISPC/
France	Yeast Structural Genomics (YSG)	http://genomics.eu.org/spip/index.php
	Bacterial Targets at IGS-CNRS (BIGS)	http://igs-server.cnrs-mrs.fr/Str_gen/

structures and tested the feasibility of a high-throughput structure production pipeline. PSI1 proved to be very productive, with more than 1100 structures solved over the five-year period, illustrating the immense potential for expediting protein structure solution through focused investments. The new technologies pioneered have already found their applications in conventional structural biology laboratories to facilitate the structural characterization of more difficult targets. The results and progress of major structural genomics initiatives in the U.S.A. and around the world have been recently summarized (2).

In July, 2005, the PSI advanced into a production phase. PSI2 consists of two major components: large-scale centers to increase the structural coverage of sequenced genomes by high-throughput production of structures and specialized centers to reduce technical barriers to high-throughput structure solution of challenging proteins (such as integral membrane proteins and multi-protein complexes). In addition to the production centers, centralized databases are being set up to coordinate the target selection from each center and to disseminate results to the public (Target Search for Structural Genomics (TARGETDB) at http://targetdb.pdb.org/ and Protein Expression Purification and Crystallization Database (PEPCDB) at http://pepcdb.pdb.org/). The objective of PSI2 is to solve 3000-4000 protein structures in a five-year period at a cost of ~$50,000-75,000 per structure and efficiently fill in the gaps in protein 'fold space' from all kingdoms of life. The large influx of the protein structures will benefit all structural biologists and other scientific communities, and ultimately be used to assist in drug discovery. The consortia selected for PSI2 are listed in Table 2.

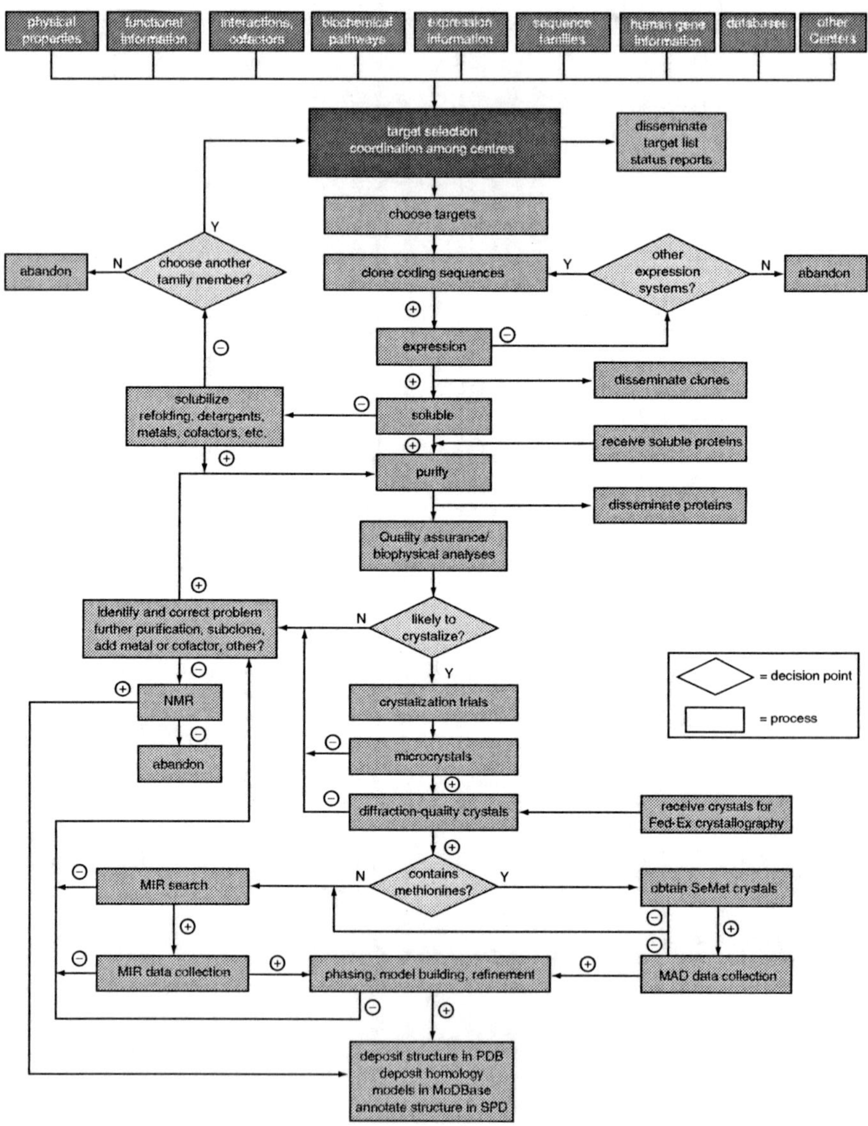

Figure 2. Schematic flow diagram of the NYSGXRC high-throughput strategy. NYSGXRC is a collaborative industrial / academic research consortium devoted to large production of protein structures and is one of the large scale centers selected for PSI2.

The New York Structural GenomiX Research Consortium (NYSGXRC), a collaborative industrial/academic research consortium devoted to large-scale production of protein structures, is one of the large-scale centers funded within PSI2. This review will focus on methodologies and technologies developed by NYSGXRC for high-throughput protein structure determination, as well as those contributed from other PSI Centers. Within NYSGXRC, proteins purified by a biotechnology/pharmacology company, SGX Pharmaceuticals Inc.

Table 2. Protein Structure Initiative (PSI)—2 in USA.

Consortium	Led By	Focus / Web Address
NIH Affiliated Large-Scale Protein Structure Production Centers		
Joint Centre for Structural Genomics JCSG	Ian Wilson Scripps Research Institute La Jolla, CA	Novel cell signaling proteins from *C. elegans*, *human*, *mouse* and *Drosophila*. http://www.jcsg.org/
Midwest Center for Structural Genomics MCSG	Andrzej Joachimiak Argonne National Laboratory Near Chicago, IL	Plan to solve quickly large number of "easy" targets through highly cost-effective methods http://www.mcsg.anl.gov/
New York Structural GenomiX Research Consortium NYSGXRC	Stephen Burley Structural GenomiX Pharmaceuticals San Diego, CA	Novel folds and biologically important proteins from three kingdoms of life. http://www.nysgxrc.org/
Northeast Structural Genomics Consortium NESGC	Gaetano Montelione Rutgers University New Brunswick, NJ	Eukaryotic model organisms which are subjects of extensive functional genomics research, including *S. cerevisiae*, *C. elegans* and *D. melanogaster*, as well as homologs from the human genome. http://www.nesg.org/
NIH Affiliated Specialized Centers		
Accelerated Technologies Center for Gene to 3D Structure	Lance Stewart deCODE biostructures Bainbridge Island, WA	Development, operation and deployment of novel approaches in miniaturization, integration and automation with an aim towards lowering the overall cost of gene to structure. http://www.atcg3d.org
Center for Eukaryotic Structural Genomics	John Markley University of Wisconsin Madison, WI	NMR spectroscopy and its biological applications; structure function relationships in proteins. http://www.uwstructural genomics.org/
Center for High-Throughput Structural Biology	George De Titta Hauptman-Woodward Medical Research Institute Buffalo, NY	Development of crystal growth methods and techniques. High-throughput structural biology. Website: forthcoming.
Center for Structures of Membrane Proteins	Robert Stroud University of California San Francisco, CA	Large effort to express eukaryotic membrane proteins with the end goal of determining their molecular structures. http://csmp.ucsf.edu/index.htm
Integrated Center for Structure and Function Innovation ISFI	Thomas Terwilliger Los Alamos National Laboratory Los Alamos, NM	Powerful methods for screening whether a molecule is properly folded and whether it will crystallize. Website : forthcoming.
New York Consortium on Membrane Protein Structure	Wayne Hendrickson New York Structural Biology Center New York, NY	Key class of proteins that serve as the portals through which cells and some components within cells communicate with the external environment. Membrane proteins lead to the development of disease and many are pharmaceutical targets of prime interest. http://www.nysbc.org/

(SGX Pharma), are distributed to four academic institutions: Albert Einstein College of Medicine (AECOM), Brookhaven National Laboratory (BNL), Case Western Reserve University (CWRU), and Columbia University (CU), for structural studies. Bioinformatics, target selection and data management tasks are performed by AECOM and University of California San Francisco (UCSF) (Table 3).

TARGET SELECTION

During PSI1, NYSGXRC and other structural genomics centers independently developed strategies for target selection. In NYSGXRC, targets were selected from microbes to human, with particular emphasis on proteins of biomedical relevance and 'hypothetical' proteins with unknown function. Because of their

Table 3. NYSGXRC Scientific Organization.

Organization Name	Scientific Team Leader	Tasks
SGX Pharmaceuticals Inc. SGX Pharma	Stephen K Burley *Principal Invstigator (PI)*	Protein Production
Albert Einstein College of Medicine AECOM	Steven Almo Institutional *Co-PI* *Department of Biochemistry* almo@aecom.yu.edu	Protein Structure Determination
	Andras Fiser *Co-PI* *Department of Biochemistry Center for Bioinformatics* fiser@fiserlab.org	Target selection, Data management and functional annotation
Case Western Reserve University CWRU	Mark R. Chance *Institutional Co-PI* *Case Center for Proteomics* mark.chance@case.edu	Metalloproteomics, Protein Structure Annotation and Publication
Columbia University CU	Lawrence Shapiro *Institutional Co-PI* *Department of Biophysics* shapiro@convex.hhmi.columbia.edu	Protein Structure Determination
University of California San Francisco UCSF	Andrej Sali *Institutional Co-PI* *California Institute for Quantitative Biomedical Research* sali@salilab.org	Comparative modeling
Brookhaven National Laboratory BNL	S. Swaminathan *Institutional Co-PI* *Biology Department,* swami@bnl.gov	Protein Structure Determination
	F. William Studier *Co-PI* Biology Department studier@bnl.gov	Protein expression strategies

biological importance and relative ease to work with, enzymes associated with small molecule metabolic pathways were also frequently selected. The target proteins were chosen based on their low sequence homology to the proteins with known structures. In addition, orthologues from several species were simultaneously cloned and purified to maximize the chance of solving the fold. Once the representative structure is solved, the efforts to solve the other orthologues are abandoned.

In PSI2, the majority of targets will be chosen using a centralized strategy as imposed by NIH target selection committee (http://grants2.nih.gov/grants/guide/rfa-files/RFA-GM-05-001.html) (3). Several strategies have been suggested and discussed in detail (4, 5) such as the "Pfam5000" strategy, which involves selecting the 5,000 largest families from the Pfam database as sources for targets. It is estimated that if at least one structure is solved from each of these 5,000 families, it will provide sequence coverage of 68% of prokaryotic proteins and 61% of eukaryotic proteins, and greatly increase our ability to assign folds for all sequenced genomes through modeling and threading methods. Pfam5000 strategy complements the other strategies such as random target selection strategies and single-genome strategies (5). In PSI2, NIH requests a target selection strategy that combines coarse-grained coverage of sequence space, proteins of known medical interest, and contributions from the scientific community (5). NYSGXRC will follow the method suggested by NIH target selection committee. About 70% of the targets will be selected from available genomes in coordination with the other 3 large scale structural genomics centers in PSI2.

PROTEIN PRODUCTION AND BIOPHYSICAL ANALYSIS

During PSI1, SGX Pharmaceuticals Inc, (http://www.sgxpharma.com/) established a modular industrial platform for the recombinant protein production. cDNAs of interest were cloned by PCR amplification and inserted into a suitable expression vector. The procedure can be operated in a parallel fashion for high-throughput (6). The protocol developed by NYSGXRC laboratories, using the T7 RNA polymerase-dependent *E.coli* expression vector system (pET-vectors), is a universal system to generate recombinant protein for structural analysis (7, 8). pET vectors are usually combined with *E. coli* B strain BL21 or the derivatives that are engineered to carry the T7 RNA polymerase gene. These strains, however, have limitations in cloning and stable propagation of the expression constructs. The approach based on the concept of topoisomerase mediation, which involves directional flap ligation of a blunt-ended PCR product into pET100/D-TOPO Vector (Invitrogen) was adopted by NYSGXRC. It creates a fusion protein bearing an N-terminal His_6-tag followed by a polio viral protease cleavage site followed by the protein sequence of interest. Recently, NYSGXRC implemented an additional vector for recombinant protein expression based on N-terminal fusions with a yeast form of SUMO, a small ubiquitin-like modifier that frequently enhances the solubility to the recombinant fusion protein (9, 10). The pSUMO system utilizes an N-terminal His_6-tag SUMO fusion with the respective target sequence. The protein is expressed in bacteria, purified by metal affinity chromatography, and liberated from the His_6-SUMO fusion by cleavage with a modified version of the desumoylating enzyme Ulp1.

To facilitate the high-throughput production of proteins, a Beckman Biomek FX robotic platform has been adopted to perform many of the steps required from PCR to transformation in 96-well format with bar code tracking of sample and reagent plates (6). Some steps are conducted off-line with multichannel pipetting. Small-scale (1 μg) purification of recombinant proteins followed by spotting onto a matrix-assisted laser desorption ionization mass spectrometry (MALDI-MS) sample plate allows rapid identification of constructs expressing the appropriate product (11).

All soluble proteins purified were subjected to biophysical analyses, including mass spectrometry for construct verification and protein purification, analytical gel filtration for homogeneity, domain mapping by limited proteolysis combined with mass spectrometry (LPMS) to analyze for "floppy-ends", peptide mapping of posttranslational modifications *via* mass spectrometry, and UV/vis. absorbance spectroscopy to identify possible bound co-factors (6).

Proteins produced at SGX Pharmaceuticals (10+ mg) were shipped to AECOM, BNL and CU for protein structure determination (Table 3) and smaller amounts of samples (0.1 mg) were also shipped to Case Center for Proteomics, CWRU for intrinsic metal detection through automated X-ray absorption spectroscopy (XAS) at the NSLS beamline X9B (12, 13).

The results of these measurements are made available to the crystallographers to facilitate *de novo* phasing and structure solution using the intrinsic metals (14). SGX Pharmaceuticals provides selenomethionine (SeMet)-labeled proteins for X-ray single/multiple anomalous dispersion (SAD/MAD) studies at synchrotron sites after adequate crystallization conditions have been established for native crystals. The use of synchrotron radiation is crucial to the NYSGXRC pipeline. High brilliance and energy tuneability (mainly Se Edge) are prerequisites for fast data collection from small protein crystals (15).

PROTEIN STRUCTURE ANALYSIS

X-ray crystallography is the primary technique used for protein structure determination in most of structural genomics centers. The efforts to solve the protein structures have seen great improvements over the past decade and resulted in dramatic accumulation of protein structures in the PDB (http://www.rcsb.org/pdb).

Crystallization

It is usually regarded as a major bottleneck for structure determination by X-ray crystallography with low success rates evident in all structural genomics centers. As of October 2005, for all structural genomics centers worldwide, only 4,692 targets (8% of the proteins cloned) yielded crystals and 2,034 targets (3.5%) resulted in crystal structures. In PSI centers, 3,629 targets (7% of the proteins cloned) were crystallized yielding 1,210 (2%) crystal structures (as of October 2005). For structural genomics centers focusing on medically related human proteins, the statistics are even lower. For example, within the Protein Structure Factory (Germany), only 63 targets (7% of the proteins cloned) were crystallized and 11 (1.2%) protein structures were solved. Frequently, the failure of producing diffracting quality crystals is attributed to disordered regions, particularly at

N- and C-termini. These full length proteins have high tendency to aggregate with low yield (2 mg/L) and are difficult to concentrate (>1 mg/ml). Several techniques have been developed to identify the stable domains and to remove the structural micro-heterogeneity of the proteins. The technique, termed limited proteolysis by mass spectroscopy (LPMS) (16, 17), has been implemented in the NYSGXRC pipeline (18). The constructs obtained by LPMS possess enhanced qualities such as high yield (30 mg/L), stability overtime and greater tendency to crystallize (19). It has been demonstrated that the targets resistant to proteolysis are good candidates for crystallographic studies with 27% of these targets yielding 3-dimensional structures thus far, whereas only 9% of the targets showing partial proteolysis yielded 3-dimensional structures to date. Large-scale sub-cloning and subsequent testing of expression, solubility, and crystallization are currently underway (18). Another technique, the hydrogen/deuterium exchange mass spectroscopy (DXMS) that allows rapid identification of unstructured regions in proteins, was developed at JCSG (20). These targets, TM0160, TM1171, TM0613 and TM0021, have been successfully crystallized after DXMS analysis (21). Bioinformatics strategies can also be applied to identify disorder region using computational tools such as *PONDR* (22), *GlobPlot* (23) and *DisEMBL* (24), and to help redesign constructs utilizing different expression systems and genes from different species. For example, crystals suitable for X-ray crystallographic studies for Protein Structure Factory (Germany) target PSF200001226 (PDB ID 1U2H) were obtained following the truncation of the first 14 N-terminal amino acids that were predicted to be structurally disordered and these crystals diffracted to 0.96 Å resolution (25).

The field of protein crystallization is revolutionized with the development of robotic technology. All steps in crystallization have been automated including crystallization plate setup and bar coding, movement of crystallization plates into and out of the storage vault, and crystallization plate imaging, image processing, storage and display. These traditionally tedious manual procedures have been addressed to save proteins, time and cost of the crystallization experiments. The robotic imaging systems (26) and macromolecular crystallization using free-interface diffusion method at the nanoliter scale (27) have been described recently. The system is capable of performing multidimensional screening (mixing 5-10 solutions) to explore more crystallization space, maximizing the chance of obtaining crystals. Furthermore, capillary-containing protein crystals can be directly mounted on the goniometer, eliminating the need of crystal manipulation and mounting. Currently, NYSGXRC has implemented parallel robotic stations for high-throughput crystallization screening at each crystallography site utilizing 96-well "sitting drop" vapor diffusion method. The optimization screens are still performed manually.

Synchrotron Data Collection

New third generation synchrotrons with beamlines equipped with insertion devices provide more intense, tunable and stable X-ray beams, allowing crystallographers to collect higher quality data much more rapidly. As of August 2005, beamline 19ID-APS (183 deposits) has the highest number of deposited PDBs among the beamlines utilized by PSI centers. Several bending magnet beamlines including X4A (90 deposits), 19BM-APS (88 deposits) and X9A-NSLS (53 deposits) are also making significant contributions. Other factors such as flash-freezing techniques, faster and larger CCD X-ray detectors have led to dramatic increases in the rate of structure determination. Novel methods for automatic crystal mounting, optical crystal centering, data collection and indexing of the crystals have been developed at many synchrotron sites (28). During PSI1, NYSGXRC built highly collimated and extremely intense beamline, X29-NSLS, a novel mini-gap undulator beamline, for efficient high-resolution data collection from very small crystals to facilitate rapid structure determination (29). The X29 optical system comprises a double crystal monochromator with a sagittally bent second crystal providing horizontal focusing, followed by a cylindrically bent mirror providing vertical focusing and harmonics rejection. The photon energy range of the monochromator is 4-18 keV which covers the absorption edges of all commonly used heavy atoms (30). The method of MAD phasing requires X-ray diffraction measurements at two to four X-ray energies near an atomic absorption edge of the heavy atom, chosen to maximize the real and imaginary components of anomalous scattering. MAD phasing on data collected from crystals containing variety of anomalous scatterers including Se, Fe, Cu, Br, Tb, Pt, Hg, W, Au and Zn, is the method of choice for determining new crystal structures. In addition, X29 is equipped with state-of-the-art ADSC Q315 detector system (near 100 μm resolution with near 2 second readout time) in order to take advantage of the short exposure time (~5 sec per frame) and to collect data on the crystals with large unit cells (>600 Å). Furthermore, the X29 station is equipped with gaseous liquid-nitrogen cooling, highly automated beamline control, efficient software packages to facilitate high-throughput data collection. Installation of sample changing and crystal alignment robotics which automate the initial crystal screening step are underway at X29.

The availability of powerful computers contributes to high speed data collection by automation in selection of optimum data acquisition parameters and processing protocols (31). As of August 2005, the program HKL was used on fly for integration (716 PSI deposits) and for scaling (743 PSI deposits). Another other popular program for data integration is MOSFLM (154 PSI deposits) and for data scaling is SCALA (174 PSI deposits). However, since protein crystals differ enormously in their diffraction properties, it is difficult to develop a complete automated system using a single data collection strategy that can satisfy all possible scenarios (32).

Phasing

One of the primary problems in macromolecular X-ray crystallography is the phase problem. Single/multiple anomalous dispersion (SAD/MAD), single/multiple isomorphous replacement (SIR/MIR) and molecular replacement (MR) methods are commonly used to solve the phase problem. Recently, phasing using SAD/MAD with SeMet-substituted proteins has become a routine process in protein crystallography (33). In 2004, the percentage of newly-deposited structures, which share less than 30% of sequence identity to any known structures at the time of deposition, are 61% (555 of 915), 63% (326 of 521) and 77%

(62 of 82), respectively, for all SG (Structural Genomics) centers, PSI centers and NYSGXRC. These statistics indicate that the majority of protein structures from Structural Genomics centers are determined using SAD or MAD methods. Novel methods, such as heavy atom derivatization with halides, SAD with sulfur atoms, phasing using Hg radiation damage and brute force molecular replacement, were developed by NYSGXRC to facilitate high-throughput structure determination (34-36).

Automated Structure Solution

Development of integration and extension of existing crystallographic software provides user-friendly tools for rapid automated structure determination. Several integrated program packages are now in general use and listed in Table 4. Two popular automated protocols are commonly used by the NYSGXRC. First, the program *HKL2MAP* connects *SHELX* suite (37). The processed data from HKL-2000 collected at different wavelengths are scaled for data analysis, phase calculation and the electron density map are displayed using XFIT (38). The electron density map can be interpreted and fitted through automatic model building program such as *ARP/wARP* (39). Second, *SOLVE/RESOLVE* suit is fully automated and can function with data resolution as low as 3 Å (40). With these approaches, initial models can be built and displayed while the data are still being collected. With fast computers and the automated crystallographic software, structure solution is straightforward in many cases. The initial model built with automated programs is further completed with manual model fitting using programs O (41) or COOT (42), and subjected to refinement with programs REFMAC and/or CNS (43). As of August 2005, 469 PSI deposits are refined using CNS and 459 deposits are refined with REFMAC programs. Web-based tools such as AutoDep are available to deposit the coordinates and structure factors to the PDB with immediate release. Attempts are underway around the world including NYSGXRC to build user-friendly automated tools for protein structure determination (44-50).

Project Management Systems and Progress Report

An important feature of structural genomics is the on-line documentation of progress that allows data mining for evaluating the enterprise. Status information for all the steps of the high-throughput pipeline are archived through a centralized NYSGXRC database (http://www.nysgxrc.org) and linked to the Protein Data Bank (PDB) through the target database (http://targetdb.rcsb.org). The progress of all NYSGXRC targets is shown in Table 5. As of August, 2005, 190 protein structures have been determined in NYSGXRC. So far 11.2% of the cloned targets have yielded deposited structures (1,685 cloned:190 structures in PDB). However, this success rate is well above the 4.4% success rate indicated in the target database for all structural genomics centers worldwide (56,146 cloned: 2475 structures in PDB as of October, 2005) and the 2.3% success rate for all PSI centers (51,131 cloned: 1180 structures in PDB). Interestingly, 64% of cloned NYSGXRC targets were purified and 18% of purified proteins yielded crystal structures.

Table 4. Selected Computational Resources for Protein Structure Analysis.

Programs	Features	Web Address
DENZO/HKL-2000	Analysis and process X-ray data collected from single crystals	http://www.hkl-xray.com/
MOSFLM	Analysis and process X-ray data collected on the image plate and CCD	http://www.mrc-lmb.cam.ac.uk/harry/mosflm/
HKL2MAP/SHELX	Direct methods, phasing refinement and heavy atoms	http://shelx.uni-ac.gwdg.de/SHELX/
SnB	Direct methods, phasing refinement and heavy atoms	http://www.hwi.buffalo.edu/SnB/
BnP	Substructure determination and refinement	http://www.hwi.buffalo.edu/BnP/
SOLVE/RESOLVE	Heavy atom refinement, phasing and chain tracing	http://www.solve.lanl.gov/
AUTOSHARP	Statistical heavy atom refinement and phasing	http://www.globalphasing.com/sharp/
PHENIX	Python-based Hierarchical Environment for Integrated Xtallography	http://www.phenix-online.org/
CCP4	Collaborative Computing Project 4	http://www.ccp4.ac.uk/main.html
ARP/wARP	Automatic model refinement and chain tracing	http://www.embl-hamburg.de/ARP/
CNS	Refinement, phasing, heavy atoms, molecular replacement	http://asdp.bio.bnl.gov/cns_solve_1.1/doc/html/index.html
O	Graphics—model and map visualization and building	
XtalView	Solving and building crystal structures	http://www.sdsc.edu/CCMS/Packages/ XTALVIEW/xtalview.html
COOT	Crystallographic Object-Oriented Toolkit. Model building and validation.	http://www.ysbl.york.ac.uk/~emsley/coot/
PyMol	A molecular graphics system with an embedded Python interpreter	http://pymol.sourceforge.net/
USF	Uppsala Software Factory—supportive programs	http://alpha2.bmc.uu.se/usf/
PROCHECK	Check stereochemistry quality of protein models	http://www.biochem.ucl.ac.uk/~roman/ procheck/procheck.html
SFCHECK	Assessment between Structure Factors and models	http://www.ysbl.york.ac.uk/~alexei/sfcheck.html
AutoDep	Model validation and deposition.	http://deposit.rcsb.org/adit/ http://www.ebi.ac.uk/msd-srv/autodep4/index.jsp

Table 5. Progress of NYSGXRC as of August 16, 2005*.

Different Stages of the Pipeline	
Targets Selected	2306
Cloned	1685
Expressed	1375
Soluble	1164
Purified	1068
Crystallized	391
Diffraction Quality Crystals	240
Protein Structures in PDB	190
Protein Structures Released by year	
<2000	6
2001	12
2002	14
2003	35
2004	82
2005	41

*Updates available at http://www.nysgxrc.org and mirrored http://targetdb.pdb.org/statistics/sites/NYSGRC.html.

In contrast, only 19% of the cloned targets from all PSI centers were purified and 13% of purified proteins resulted in crystal structures. The statistics are similar for all SG centers where only 21% of cloned targets were purified and 17% of purified proteins resulted in crystal structures. The growth of protein structures released by NYSGXRC by year is shown in Table 5. The number of protein structures released in 2004 by all SG and PSI centers (including NYSGXRC) are 914 and 523, respectively. The contribution from NYSGXRC alone (82 structures for 2004) is about 9.7% of structures by all SG centers and 18% by PSI centers.

The quantity of structures solved may not be the best measure and it is important to analyze the quality of the structures solved within the various projects. The productivity depends upon the nature of targets and the availability of high-throughput methodologies and the technical infrastructure to tackle them. Several beamlines at National Synchrotron Light Source (NSLS) and Advanced Photon Source (APS) have been utilized for X-ray diffraction data collection. The majority of structures were determined from the data collected at NSLS X9A (52 structures) and APS 31ID (33 structures). In addition, 19 structures were from NSLS X29 (29). The average resolution of all target structures by the consortium is 2.26Å. The average R_{work} and R_{free} for these structures are 0.211 and 0.25, respectively, indicating the high quality of the structures determined by NYSGXRC. The average sequence length for NYSGXRC deposited structures is 296 residues (as of August 2005), significantly higher than the average by all PSI centers (358 residues). For the other three large PSI centers selected for the production phase, the average lengths are 211, 177 and 255, respectively, for MCSG, NESGC and JCSG. The range of organisms from which the NYSGXRC targets were selected from reflects the broad focus of the NYSGXRC (Table 6). As of August, 2005, 149 structures are from prokaryotes (archaea and bacteria) and 32

Table 6. NYSGXRC Structures by Organism (as of August 16, 2005*).

Organisms from Three Kingdoms of Life	No. of organisms × structures	Total
Escherichia coli	1 × 37	37
Bacillus subtilis	1 × 16	16
Saccharomyces cerevisiae	1 × 13	13
Enterococcus faecalis; Pseudomonas aeruginosa;	2 × 8	16
Agrobacterium tumefaciens; Haemophilus influenzae; Methanococcus jannaschii; Mus Musculus; Vibrio cholerae;	5 × 5	25
Archaeoglobus fulgidus; Deinococcus radiodurans; Homo sapiens; Thermotoga maritima	4 × 4	16
Bacillus halodurans; Mycobacterium tuberculosis; Neisseria meningitides; Streptococcus pneumoniae; Streptococcus pyogenes;	5 × 3	15
Campylobacter jejuni; Clostridium acetobutylicum ATCC 824; Listeria monocytogenes; Phleum pretense; Salmonella typhimurium; Schizosaccharomyces pombe; Staphylococcus aureus.	7 × 2	14
Aquifex aeolicus; Arabidopsis thaliana; Bacteroides thetaiotaomicron; Borrelia burgdorferi; Bradyrhizobium japonicum; Caenorhabditis elegans; Caulobacter crescentus; Chlorobium tepidum TLS; Encephalitozoon cuniculi; Helicobacter pylori J99; Klebsiella pneumoniae; Listeria innocua Clip11262; Salmonella enterica; Shigella flexneri; Streptococcus mutans UA159; Thermoplasma acidophilum; Xanthomonas campestris	17 × 1	17

*Updates available at http://www.nysgxrc.org and mirrored http://targetdb.pdb.org/statistics/sites/NYSGRC.html.

structures are from eukaryotes (yeast, plasmodium, arabidopsis, nematode, fly, mouse, humans). The statistics indicate that NYSGXRC has been very productive. The number of high-quality crystal structures by NYSGXRC through the high-throughput pipeline is promising, however, it could also be argued that many of these structures are "easy" targets (low hanging fruit), so that both quantity and quality are expected to be high. More challenging targets, such as human and other eukaryotic proteins, and large macromolecule assemblies, pose a greater challenge. The functional coverage of NYSGXRC structures based on enzyme classification, biological process, cell component, molecular function and disease is shown in Table 7. About 37% of the solved structures are hypothetical proteins with unknown function.

In order to annotate proteins with unknown function, high-throughput tools are needed at each step of the experimental pipeline, including the timely release of protein structures to biologists and other scientists. For instance, out of 190 protein structures by NYSGXRC to date, 140 (74%) of them are "to be published" and only about 50 (26%) have peer-reviewed publications. Similar statistics can also be found for other structural genomics centers around the world.

Table 7. NYSGXRC Structures by Functional Classification (as of August 16, 2005).

Classification	**Functional coverage × structures**	**Total**
Unknown Function	1 × 62	62
Transferase	1 × 23	23
Hydrolase	1 × 16	16
Oxidoreductase	1 × 14	14
Lyase, Isomerase	2 × 8	16
Transcription	1 × 7	7
DNA Binding	2 × 4	8
Structural Protein	1 × 3	3
Signaling; Protein Binding; Lipid Binding; Ligase; Harmone/Growth Factor; Biosynthetic; Allergen	7 × 2	14
Penicillin Binding; Immune System	2 × 1	2

HOMOLOGY MODELING OF REPRESENTATIVE PROTEIN FAMILY MEMBERS

Homology modeling or comparative modeling takes advantage of structural similarities within protein families. This technique is based on the assumption that all the homologous members of the protein family are related by divergent evolution from a common ancestor and must share a common basic fold. Solving the structure of any single member of a protein family clustered at 30% or more identity allows comparative modeling of the entire family in most cases. Basic approaches to homology modeling were initiated by Greer in 1981 (51) and Sali and Blundell in 1993 (52), and the methods were recently reviewed (53). Automated homology modeling with MODWEB has been fully implemented by NYSGXRC and is now being used routinely by NYSGXRC members, other PSI centers and researchers around the world (54). About 146,236 protein structure models including 12,651 accurate models have been generated using 181 NYSGXRC structures with an average number of models per structure of 807. The quality and usefulness of homology models depend critically on the level of sequence identity. The accuracy of a model based on a template with >50% sequence identity is equal to a medium-resolution crystal structure (3.0 Å resolution). Models based on >30% but less than 50% sequence identity are suitable for many applications including fold assignment and molecular replacement for phasing. Models based on <30% identity have the possibility of significant alignment errors. As a general rule, 30% sequence identity is the arbitrary cutoff for effective homology modeling. Despite the possible errors, less accurate models are useful in many applications in structural biology. For examples, they can be used to identify putative active site residues, redesign expression constructs, and as templates for structure determination by molecular replacement. Recently, combining homology modeling results with low resolution electron microscopic maps have been shown to help model more difficult targets, such as macromolecular complexes and eukaryotic proteins, and this approach is becoming well accepted (55-57).

STRUCTURE TO FUNCTION

Several bioinformatics servers, such as ProFunc (58) and ProTarget (59) servers, have been developed and are available for public access for protein structure and sequence analysis which includes prediction of the function of proteins from the solved structures. The information available from the three-dimensional structure of a protein, relating to its function, is summarized in Figure 3a. The theory and practice of how to predict function from sequence and structure have been thoroughly discussed (60-63). Currently, there are two sequence-based approaches for protein annotation. Enzyme classification of enzymes (EC numbers) has been used to study the sequence, structure and function relationships (64-66). Second, the Gene Ontology (GO) provides a consistent view of molecular function, biological process and cell component beyond enzymes (67).

The summary of biological function information extracted from the protein structures and structure-based function discovery has been described (57, 68). One of the most powerful methods for functional inference is identification of homologous proteins and protein structures through structural comparisons (69, 70). The proteins can diverge beyond significant sequence similarity but still retain the 3D fold of their ancestor and even similar functions. Commonly used web-based servers to scan the novel protein structure against the known protein structure database (PDB) and retrieve closest matches are DALI (71) and VAST (72). For example, the crystal structure of *E.coli* L-Arabinose isomerase (NYSGXRC target T2031, PDB ID 2AJT) shows significant similarity to that of *E.coli* fucose isomerase (PDB ID 1FUI) despite the very low sequence identity (9.7%) shared by the two enzymes (Figure 3b). Both structures retain hexameric subunit assembly for enzyme activity based on the results from electron microscopy studies (73). In addition, the two enzymes show similar substrate specificities (74).

However, DALI and VAST servers usually fail to produce any match if the new protein structure possesses a novel fold. In this case, identification of functional sites across different folds is required (75). Using databases of active site templates, programs such as PINTS (76), PROCAT (77) and Rigor (78) identify conservation of functional patterns within the structures of different folds. If both structure comparison and functional site comparison for proteins with unknown function fail to yield any match, analysis of the sequence conservation through evolution may reveal their functions. Conservation score can be calculated for each residue in the sequence by comparing the residue variability at each position in a multiple sequence alignment of homologous proteins, and mapped onto the protein surface. The web-based server ConSurf (79, 80) identifies most likely functional or protein-protein interaction patches on the surface of the protein structure. Further, analysis of clefts and cavities on the protein surface can be useful to locate its putative active site and sometimes provide clues to its function. The program SURFNET (81) performs the analysis of clefts and their surface properties automatically and points to the regions that are most likely to be functionally important. Many of the bioinformatics approaches described above are routinely in use to annotate protein functions.

Presence of metal atoms in proteins often marks active centers and provides a guide to functional annotation. For example, the crystal structure of ybeY protein from *E.coli* (NYSGXRC target T842, PDB ID 1XM5) reveals that the

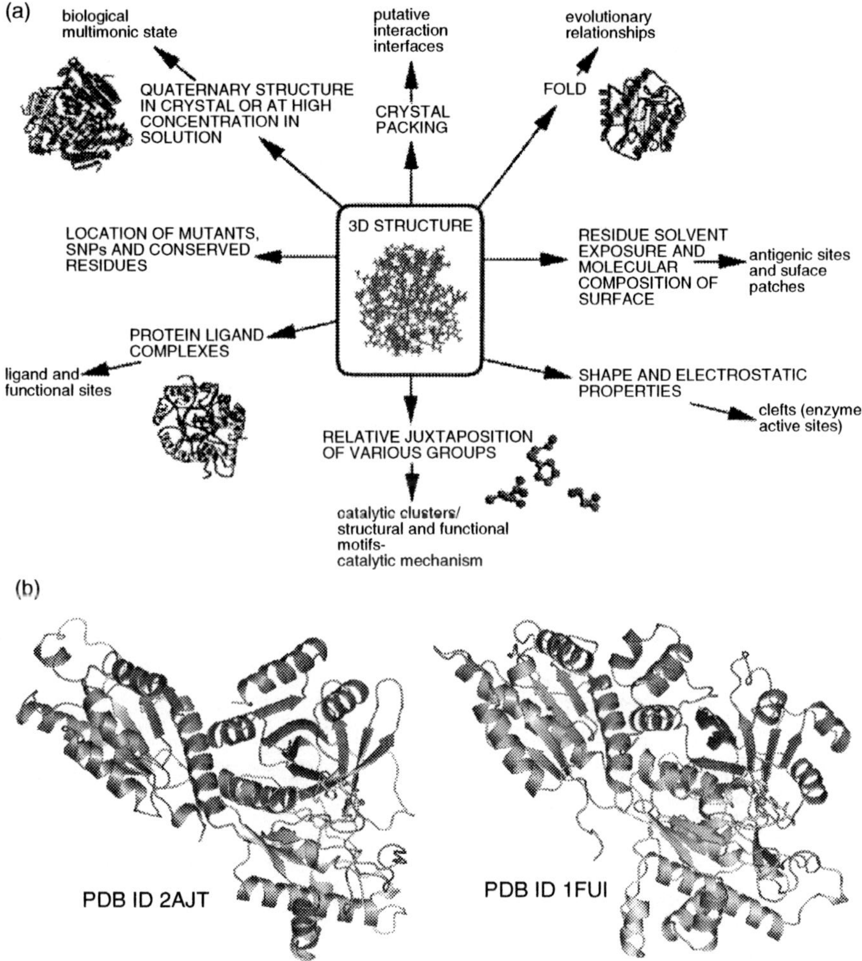

Figure 3. Structure to function and examples. (a) Summary of information derived from protein structure, with biological function related. Taken from Thornton, J.M., Todd, A.E., Milburn, D., Borkakoti, N., Orengo, C.A. (2000) From Structure to function: approaches and limitations. Nat. Struct. Biol. 7(Suppl.), 991-994. (b) Structural conservation in distant evolutionary relatives, *E.coli* L-Arabinose isomerase (PDB ID 2AJT, left) and *E.coli* fucose isomerase (PDB ID 1FUI, right), in the absence of significant sequence identity; and

protein binds to a metal ion in a tetrahedral geometry with three histidine residues (Figure 3c). The fourth coordination site might be a water molecule which was not seen in the structure. The structure of ybeY and its sequence similarity to a number of predicted metal-dependent hydrolases suggests a potential functional assignment for this protein (82). A high-throughput technology to identify proteins containing metals has been developed based on X-ray fluorescence analysis to analyze for transition metal content at beamline X9B of National Synchrotron Light Source. The initial results and potential application towards protein annotation have been discussed recently (13).

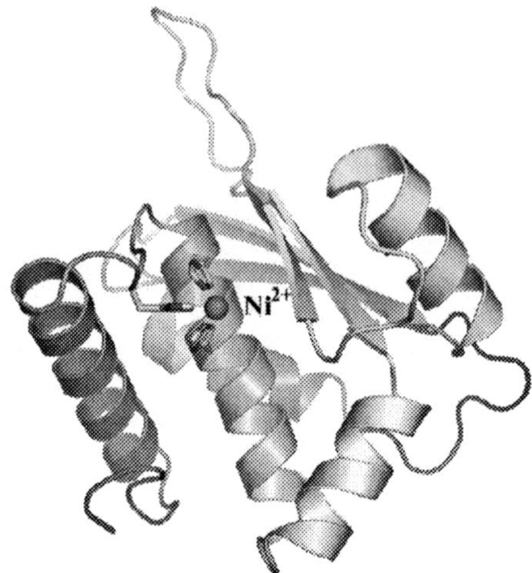

Figure 3. (*Continued*) (c) Presence of a metal ion, Ni^{2+}, guides protein annotation for NYSGXRC target T842 (PDB ID 1XM5).

The observation of an unexpected bound ligand sometimes gives clues to protein function annotation (83-85). For example, the crystal structure of the *E.coli* Ycei periplasmic protein (NYSGXRC target T792, PDB ID 1Y0G) revealed a dimer of β-barrels (similar to lipocalin superfamily folds) with a continuous electron density feature running along the entire length of the central axis of the β-barrels. The electron density was interpreted as 2-octaprenylphenol (OPP) and mass spectroscopic studies are under way to confirm the identity. The OPP bound to Ycei helps to identify the active site. In principle, experimental approaches such as functional assays and site-directed mutagenesis should follow to confirm the annotation (86, 87). High-throughput methods to automate enzymatic analysis, termed as enzyme genomics, by screening for protein–ligand complex libraries using mass spectrometry and matrix-assisted laser desorption ionization-time of flight (MALDI-TOF) have been initiated (88-90). The development of proteomics strategies for genome annotation are in progress by utilizing the structure-based functional discovery (91-93).

Auto Publish Web Tool

To speed up protein structure publications and annotation, NYSGXRC is developing an automated server to prepare structure reports automatically in short structure report format of journal Acta Crystallographica F. The web-tool aims at facilitating publication of newly-solved structures by automating major steps in data analysis and manuscript preparation, such as producing tables, figures, and performing standard-functional analysis based on structure and sequence. The server generates five outputs as shown in the flowchart (Figure 4). Users will be able to start with the desired PDB code and obtain a raw manuscript that consist all the standard requirements of a regular report on a crystallographic protein structure. First, a WORD format file useful for 'Materials and

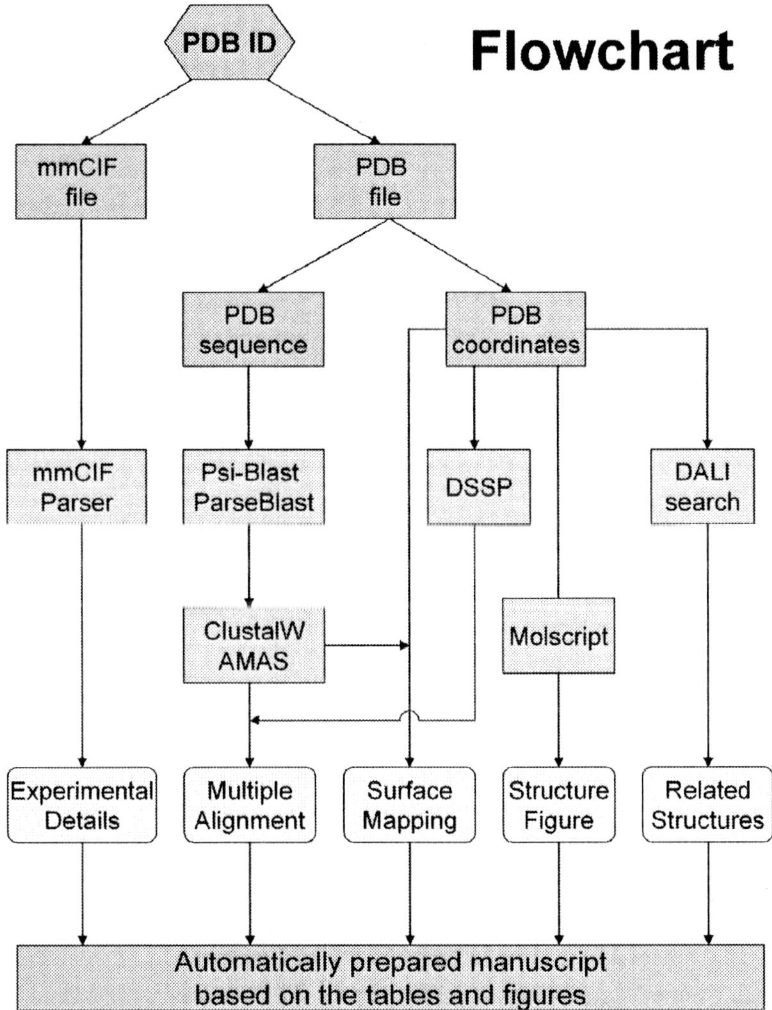

Figure 4. Flowchart of Auto Publish web server which generates five outputs including experimental details, structure images and standard functional and structural analysis to facilitate rapid publication.

Methods' section is generated along with a table containing statistics for data collection, structure solution and refinement by extracting parameters from the mmCIF PDB file. Second, the amino acid sequence of the protein structure is compared using PSI-Blast (94) with the sequences of homologs from databases and sequence conservation analysis is performed with AMAS (95) and displayed with ALSCRIPT (96) programs. The conservation scores for each residue are placed in the temperature factor column in PDB file and the modified PDB file can be automatically uploaded into graphic programs such as PyMol (97) to generate a protein surface plot with the conserved regions appropriately color coded. In addition, a standard sequence alignment figure with selected homologs is also generated and conserved residues are identified and highlighted to assist in protein family classification and functional annotation. Third, protein structure is automatically uploaded into a structure alignment program (DALI (71)). The comparative analysis results can be used for the protein fold assignment and active site identification. The web-tool has been tested on one of the NYSGXRC target structure (1XM5) (82). Currently, in-depth testing of the server is underway to prepare the structure reports more automatically by using NYSGXRC protein structures.

CONCLUSION

The NYSGXRC has implemented pipeline and potential for experimentally determining 100-200 protein structures annually. All consortium activities can be scaled up to increase capacity for protein structure production anticipated in PSI2. NYSGXRC is dedicated to unravel the shapes and to derive the functions of many hundreds of proteins in the next few years. The structural information generated in structural genomics will have a profound impact in many related fields including drug discovery by providing scientists a large structure database for structure-based drug design.

REFERENCES

1 Norvell, J.C. and Machalek, A.Z. (2000) Nat. Struct. Biol. 7 Suppl; 931.
2 Todd, A.E., Marsden, R.L., Thornton, J.M. and Orengo, C.A. (2005) J. Mol. Biol. 348(5), 1235-1260.
3 PSI-phase 1 and beyond. (2004) Nat. Struct. Mol. Biol. 11(3), 201.
4 Chandonia, J.M., Earnest, T.N. and Brenner, S.E. (2004) Genome Biol. 5, 343.
5 Chandonia, J.M. and Brenner, S.E. (2005) Proteins 58(1), 166-179.
6 Bonanno, J.B., Almo, S.C., Bresnick, A., Chance, M.R., Fiser, A,. Swaminathan, S., Jiang, J., Studier, F.W., Shapiro, L., Lima, C.D., Gaasterland, T.M., Sali, A., Bain, K., Feil, I., Gao, X., Lorimer, D., Ramos, A., Sauder, J.M., Wasserman, S.R., Emtage, S., D'Amico, K.L. and Burley, S.K. (2005) J. Struct. Funct. Genomics 6, 225-232.
7 Studier, F.W. (2005) Protein Expr. Purif. 41(1), 207-234.
8 Studier, F.W., Rosenberg, A.H., Dunn, J.J. and Dubendorff, J.W. (1990) Methods Enzymol. 185, 60-89.
9 Mossessova, E. and Lima, C.D. (2000) Mol. Cell 5(5), 865-876.
10 Reverter, D. and Lima C.D. (2005) Nature 435(7042) 687-692.
11 Huang, R.Y., Boulton, S.J., Vidal, M., Almo, S.C., Bresnick, A.R. and Chance, M.R. (2003) Biochem. Biophys. Res. Commun. 307(4), 928-934.
12 Chance, M.R., Fiser, A., Sali, A., Pieper, U., Eswar, N., Xu, G., Fajardo, J.E., Radhakannan, T. and Marinkovic, N. (2004) Genome Res. 14(10B), 2145-2154.
13 Shi, W., Zhan, C., Ignatov, A., Manjasetty, B.A., Marinkovic, N., Sullivan, M., Huang, R. and Chance, M.R. (2005) Structure (Cambr.) 13(10), 1473-1486.

14 Rajashankar, K.R., Chance, M.R., Burley, S.K., Jiang, J., Almo, S.C., Bresnick, A., Dodatko, T., Huang, R., He, G., Chen, H., Sullivan, M., Toomey, J., Thirumuruhan, R.A., Franklin, W.A., Sali, A., Pieper, U., Eswar, N., Ilyin, V. and McMahan, L. (2002) NSLS Activity Report 2, 28-32.

15 Kim, S.H. (1998) Nat. Struct. Biol. 5 Suppl; 643-645.

16 Nieves-Alicea, R., Focia, P.J., Craig, S.P., 3rd and Eakin, A.E. (1998) Biochim. Biophys. Acta 1388(2), 500-505.

17 Sharma, S., Singh, T.P. and Bhatia, K.L. (1997) Acta Crystallogr. D. Biol. Crystallogr. 53(Pt1), 116-118.

18 Gao, X., Bain, K., Bonanno, J.B., Buchanan, M., Henderson, D., Lorimer, D., Marsh, C., Reynes, J.A., Sauder, J.M., Schwinn, K., Thai, C. and Burley, S.K. (2005) J. Struct. Funct. Genomics 6(2-3), 129-134.

19 Matsui, T., Hogetsu, K., Akao, Y., Tanaka, M., Sato, T., Kumasaka, T. and Tanaka, N. (2004) Acta Crystallogr. D. Biol. Crystallogr. 60(Pt 1), 156-159.

20 Pantazatos, D., Kim, J.S., Klock, H.E., Stevens, R.C., Wilson, I.A., Lesley, S.A. and Woods, V.L. Jr., (2004) Proc. Nat. Acad. Sci. U.S.A. 101(3), 751-756.

21 Spraggon, G., Pantazatos, D., Klock, H.E., Wilson, I.A., Woods, V.L., Jr. and Lesley, S.A. (2004) Protein Sci. 13(12), 3187-3199.

22 Romero, P., Obradovic, Z. and Dunker, A.K. (2004) Appl. Bioinformatics 3(2-3), 105-113.

23 Linding, R., Russell, R.B., Neduva, V. and Gibson, T.J. (2003) Nucl. Acids Res. 31(13), 3701-3708.

24 Iakoucheva, L.M. and Dunker, A.K. (2003) Structure (Cambr.) 11(11), 1316-1317.

25 Manjasetty, B.A., Niesen, F.H., Scheich, C., Yuette, R., Gotz, F., Behlke, J., Sievert, V., Heinemann, U. and Bussow, K. (2005) BMC Struc. Biol., 5(1), 21.

26 Mayo, C.J., Diprose, J.M., Walter, T.S., Berry, I.M., Wilson, J., Owens, R.J., Jones, E.Y., Harlos, K., Stuart, D.I. and Esnouf, R.M. (2005) Structure (Cambr.) 13(2), 175-182.

27 Segelke, B. (2005) Expert Rev. Proteomics 2(2), 165-172.

28 Muchmore, S.W., Olson, J., Jones, R., Pan, J., Blum, M., Greer, J., Merrick, S.M., Magdalinos, P. and Nienaber, V.L. (2000) Structure Fold Des. 8(12), R243-246.

29 Robinson, H.H., Shi, W., Sullivan, M., Nolan, W., Schneider, D.K., Berman, L., Lynch, D., Rock, L., Rosenbaum, G., Johnson, E., Chance, M.R. and Sweet, R.M. (2005) Synchrotron Radiation News 18(5), 27-31.

30 Ogata, C.M. (1998) Nat. Struct. Biol. 5 Suppl; 638-640.

31 Leslie, A.G., Powell, H.R., Winter, G., Svensson, O., Spruce, D., McSweeney, S., Love, D., Kinder, S., Duke, E. and Nave, C. (2002) Acta Crystallogr. D. Biol. Crystallogr. 58(Pt 11), 1924-1928.

32 Dauter, Z. (2005) Prog. Biophys. Mol. Biol. 89(2), 153-172.

33 Hendrickson, W.A. and Ogata, C.M. (1997) Methods Enzymol. 276, 494-523.

34 Boggon, T.J. and Shapiro, L. (2000) Structure Fold Des. 8(7), R143-149.

35 Strokopytov, B.V., Fedorov, A., Mahoney, N.M., Kessels, M., Drubin, D.G. and Almo, S.C. (2005) Acta Crystallogr. D. Biol. Crystallogr. 61 (Pt 3), 285-293.

36 Ramagopal, U.A., Dauter, Z., Thirumuruhan, R., Fedorov, E. and Almo, S.C. (2005) Acta Crystallogr. D. Biol. Crystallogr. 61(Pt 9), 1289-1298.

37 Schneider, T.R. and Sheldrick, G.M. (2002) Acta Crystallogr. D. Biol. Crystallogr. 58(Pt 10 Pt 2), 1772-1779.

38 McRee, D.E. (1999) J. Struct. Biol. 125(2-3), 156-165.

39 Cohen, S.X., Morris, R.J., Fernandez, F.J., Ben Jelloul, M., Kakaris, M., Parthasarathy, V., Lamzin, V.S., Kleywegt, G.J. and Perrakis, A. (2004) Acta Crystallogr. D. Biol. Crystallogr. 60(Pt 12 Pt 1), 2222-2229.

40 Terwilliger, T. (2004) J. Synchrotron Radiat. 11(Pt 1), 49-52.

41 Jones, T.A., Zou, J.Y. and Cowan, S.W. (1991) Acta Crystallogr. A. 47(Pt 2), 110-119.

42 Emsley, P. and Cowtan, K. (2004) Acta Crystallogr. D. Biol. Crystallogr. 60(Pt 12 Pt 1), 2126-2132.

43 Brunger, A.T., Adams, P.D., Clore, G.M., DeLano, W.L., Gros, P., Grosse-Kunstleve, R.W., Jiang, J.S., Kuszewski, J., Nilges, M., Pannu, N.S., Read, R.J., Rick, L.M., Simonson, T. and Warren, G.L. 1998 Acta Crystallogr. D. Biol. Crystallogr. 54(Pt 5), 905-921.

44 Ness, S.R., de Graaff, R.A., Abrahams, J.P. and Pannu, N.S. (2004) Structure (Cambr.) 12(10), 1753-1761.

45 Brunzelle, J.S., Shafaee, P., Yang, X., Weigand, S., Ren, Z. and Anderson, W.F. (2003) Acta Crystallogr. D. Biol. Crystallogr. 59(Pt 7), 1138-1144.

46 Holton, J. and Alber, T. (2004) Proc. Nat. Acad. Sci. U.S.A. 101(6), 1537-1542.

47 Liu, Z.J., Lin, D., Tempel, W., Praissman, J.L., Rose, J.P. and Wang, B.C. (2005) Acta Crystallogr. D. Biol. Crystallogr. 61(Pt 5), 520-527.

48 Panjikar, S., Parthasarathy, V., Lamzin, V.S., Weiss, M.S. and Tucker, P.A. (2005) Acta Crystallogr. D. Biol. Crystallogr. 61(Pt 4), 449-457.

49 Jiang, J.S. and Lin, Z. (2005) http://asdp.bnl.gov/. Private communications.

50 Fu, Z.Q., Rose, J.P. and Wang, B.C. (2005) Acta Crystallogr. D. Biol. Crystallogr. 61(Pt 7), 951-959.

51 Greer, J. (1981) J. Mol. Biol. 153(4), 1027-1042.

52 Sali, A. and Blundell, T.L. (1993) J. Mol. Biol. 234(3), 779-815.

53 Contreras-Moreira, B., Fitzjohn, P.W. and Bates, P.A. (2002) Appl. Bioinformatics 1(4), 177-190.

54 Pieper, U., Eswar, N., Braberg, H., Madhusudhan, M.S., Davis, F.P., Stuart, A.C., Mirkovic, N., Rossi, A., Marti-Renom, M.A., Fiser, A., Webb, B., Greenblatt, D., Huang, C.C., Ferrin, T.E. and Sali, A. (2004) Nucl. Acids Res. 32 (Database issue), D217-222.

55 Topf, M. and Sali, A. (2005) Curr. Opin. Struct. Biol. 15(5), 578-585.

56 Topf, M., Baker, M.L., John, B., Chiu, W. and Sali, A. (2005) J. Struct. Biol. 149(2), 191-203.
57 Jung, J.W. and Lee, W. (2004) J. Biochem. Mol. Biol. 37(1) 28-34.
58 Laskowski, R.A., Watson, J.D. and Thornton, J.M. (2005) Nucl. Acids Res. 33 (Web Server issue), W89-93.
59 Sasson, O. and Linial, M. (2005) Nucl. Acids Res. 33 (Web Server issue), W81-84.
60 Devos, D. and Valencia, A. (2000) Proteins 41(1), 98-107.
61 Iliopoulos, I., Tsoka, S., Andrade, M.A., Enright, A.J., Carroll, M., Poullet, P., Promponas, V., Liakopoulos, T., Palaios, G., Pasquier, C., Hamodrakas, S., Tamames, J., Yagnik, A.T., Tramontano, A., Devos, D., Blaschke, C., Valencia, A., Brett, D., Martin, D., Leroy, C., Rigoutsos, I., Sander, C. and Ouzounis, C.A. (2003) Bioinformatics 19(6), 717-726.
62 Hegyi, H. and Gerstein, M. (1999) J. Mol. Biol. 288(1), 147-164.
63 Wilson, C.A., Kreychman, J. and Gerstein, M. (2000) J. Mol. Biol. 297(1), 233-249.
64 Rost, B. (2002) J. Mol. Biol. 318(2), 595-608.
65 Tian, W. and Skolnick, J. (2003) J. Mol. Biol. 333(4), 863-882.
66 Tian, W., Arakaki, A.K. and Skolnick, J. (2004) Nucl. Acids Res. 32(21), 6226-6239.
67 Ashburner, M., Ball, C.A., Blake, J.A., Botstein, D., Butler, H., Cherry, J.M., Davis, A.P., Dolinski, K., Dwight, S.S., Eppig, J.T., Harris, M.A., Hill, D.P., Issel-Tarver, L., Kasarskis, A., Lewis, S., Matese, J.C., Richardson, J.E., Ringwald, M., Rubin, G.M. and Sherlock, G. (2000) Nat. Genet. 25(1), 25-29.
68 Thornton, J.M., Todd, A.E., Milburn, D., Borkakoti, N. and Orengo, C.A. (2000) Nat. Struct. Biol. 7 Suppl; 991-994.
69 Orengo, C.A., Todd, A.E. and Thornton, J.M. (1999) Curr. Opin. Struct. Biol. 9(3), 374-382.
70 Moult, J. and Melamud, E. (2000) Curr. Opin. Struct. Biol. 10(3), 384-389.
71 Holm, L. and Sander, C. (1993) J. Mol. Biol. 233(1), 123-138.
72 Gibrat, J.F., Madej, T. and Bryant, S.H. (1996) Curr. Opin. Struct. Biol. 6(3), 377-385.
73 Wallace, L.J., Eiserling, F.A. and Wilcox, G. (1978) J. Biol. Chem. 253(10), 3717-3720.
74 Manjasetty, B.A. and Chance, M.R. (2006) J. Mol. Biol., 360(2): 297–309.
75 Stark, A., Shkumatov, A. and Russell, R.B. (2004) Structure (Cambr.) 12(8), 1405-1412.
76 Stark, A. and Russell, R.B. (2003) Nucl. Acids Res. 31(13), 3341-3344.
77 Wallace, A.C., Laskowski, R.A. and Thornton, J.M. (1996) Protein Sci. 5(6), 1001-1013.
78 Kleywegt, G.J. (1999) J. Mol. Biol. 285(4), 1887-1897.
79 Landau, M., Mayrose, I., Rosenberg, Y., Glaser, F., Martz, E., Pupko, T. and Ben-Tal, N. (2005) Nucl. Acids Res. 33 (Web Server issue), W299-302.

80 Glaser, F., Pupko, T., Paz, I., Bell, R.E., Bechor-Shental, D., Martz, E. and Ben-Tal, N. (2003) Bioinformatics 19(1), 163-164.
81 Laskowski, R.A. (1995) J. Mol. Graph. 13(5), 323-330, 307-328.
82 Zhan, C., Fedorov, E.V., Shi, W., Ramagopal, U.A., Thirumuruhan, R., Manjasetty, B.A., Almo, S.C., Fiser, A., Chance, M.R. and Fedorov, E. (2005) Acta Crystallogr. F F61, 959-963.
83 Manjasetty, B.A., Delbruck, H., Pham, D.T., Mueller, U., Fieber-Erdmann, M., Scheich, C., Sievert, V., Bussow, K., Niesen, F.H., Weihofen, W., Loll, B., Saenger, W. and Heinemann, U. (2004) Proteins 54(4), 797-800.
84 Manjasetty, B.A., Powlowski, J. and Vrielink, A. (2003) Proc. Nat. Acad. Sci. U.S.A. 100(12), 6992-6997.
85 Turnbull, A.P., Kummel, D., Prinz, B., Holz, C., Schultchen, J., Lang, C., Niesen, F.H., Hofmann, K.P., Delbruck, H., Behlke, J., Müller, E.-C., Jarosch, E., Sommer, T. and Heinemann, U. (2005) Embo J. 24(5), 875-884.
86 Rajashankar, K.R., Bryk, R., Kniewel, R., Buglino, J.A., Nathan, C.F. and Lima, C.D. (2005) J. Biol. Chem. 280(40), 33977-33983.
87 Yang, Z., Savchenko, A., Yakunin, A., Zhang, R., Edwards, A., Arrowsmith, C. and Tong, L. (2003) J. Biol. Chem. 278(10), 8804-8808.
88 Kuznetsova, E., Proudfoot, M., Sanders, S.A., Reinking, J., Savchenko, A., Arrowsmith, C.H., Edwards, A.M., Yakunin, A.F. (2005) FEMS Microbiol. Rev. 29(2), 263-279.
89 Powell, K.D. and Fitzgerald, M.C. (2004) J. Comb. Chem. 6(2), 262-269.
90 Mathur, S., Hassel, M., Steiner, F., Hollemeyer, K. and Hartmann, R.W. (2003) J. Biomol. Screen 8(2), 136-148.
91 Yakunin, A.F., Yee, A.A., Savchenko, A., Edwards, A.M. and Arrowsmith, C.H. (2004) Curr. Opin. Chem. Biol. 8(1), 42-48.
92 Yee, A., Pardee, K., Christendat, D., Savchenko, A., Edwards, A.M. and Arrowsmith, C.H. (2003) Acc. Chem. Res. 36(3), 183-189.
93 Xie, L. and Bourne, P.E. (2005) PLoS Comput. Biol. 1(3), e31.
94 Altschul, S.F., Madden, T.L., Schaffer, A.A., Zhang, J., Zhang, Z., Miller, W. and Lipman, D.J. (1997) Nucl. Acids Res. 25(17), 3389-3402.
95 Livingstone, C.D. and Barton, G.J. (1993) Comput. Appl. Biosci. 9(6), 745-756.
96 Barton, G.J. (1993) Protein Eng. 6(1), 37-40.
97 DeLano, W.L. (2002) http://www.pymol.org.

NEW MASS-SPECTROMETRY-BASED STRATEGIES FOR LIPIDS

Giorgis Isaac[1*], Richard Jeannotte[1*], Steven Wynn Esch[2*] and Ruth Welti[1]

[1]Division of Biology
Ackert Hall
Kansas State University
Manhattan, KS 66506
[2]Mass Spectrometry Laboratory
Malott Hall
University of Kansas
Lawrence, KS 66045

ABSTRACT

In the past dozen years, many new strategies for mass-spectrometry-based analyses of lipids have been developed. Lipidomics has emerged as a comprehensive approach to analysis of lipids from biological systems, and the most-utilized lipidomics methodologies involve electrospray ionization (ESI) sources and triple quadrupole analyzers. While mass spectral techniques for lipid profiling have advanced, challenges in developing uniform data acquisition methods and in handling, storing, and analyzing mass spectral data remain. Investigation of other ionization methods, including matrix-assisted laser desorption/ionization (MALDI) and atmospheric pressure chemical ionization (APCI), has demonstrated that

*G. Isaac, R. Jeannotte and S.W. Esch are postdoctoral trainees of the Kansas Lipidomics Research Center. Each made a similar contribution to the review.

Genetic Engineering, Volume 28, Edited by J. K. Setlow

these are useful in specific applications. APCI is particularly amenable to analysis of less polar lipids, and MALDI provides a rapid technology with application for tissue imaging. Time-of-flight secondary ion mass spectrometry (TOF-SIMS) is particularly suited for imaging of tissues and cells.

INTRODUCTION

In contrast to the more modest analytical goals of traditional lipidology, the goal of a "lipidomic" approach is comprehensive analysis of lipids by high-throughput technology. Now emerging instrumental methods can be used to obtain insight into the diverse structural functions and biological activities of lipids. Only recently have technological developments in mass spectrometry (MS) made this ambitious goal achievable.

Lipid profiling, or lipidomic analyses, can lead directly to elucidation of the functions of particular genes and their protein products (e.g., 1-5). When compared to a wild-type control, the lipid profile of a mutant with altered gene expression can help identify the substrate(s) and product(s) of the enzyme encoded by the mutated gene (e.g., 1, 4, 5). Lipid mediators in cellular signaling events are especially attractive targets for investigation. High-throughput lipid profiling can be used to follow physiological changes in response to environmental stress, disease pathology, developmental cellular differentiation, and pharmacological intervention. Correlation of lipid composition with organismal, tissue-level, cellular, or subcellular differences or alterations can lead to the identification of metabolic or signaling pathways that are activated during a particular physiological response or in particular cellular events (e.g., 1, 4-9). Specific patterns of lipid species provide insight into the progression of disease states, including diabetes (10-13), cancer (14-15), Alzhemier's disease (16-17) and viral pathogenesis (8); the lipid biomarkers identified may prove useful as diagnostic or prognostic measures. Surveys of global changes in lipid composition offer unprecedented potential to establish new assays for clinical assessment and guided therapeutic strategy. New, mass-spectrometry-based imaging strategies also have strong potential to yield information about location of specific lipid species in cells and tissues.

The most popular and versatile emerging lipidomic methodologies depend on MS both to obtain detailed structural information about individual lipid molecular species and to measure simultaneously the amounts of all lipid species. Ionization and gas phase fragmentation characteristics of different lipid classes play a major role in the selectivity and sensitivity of MS (mass spectrometry) methods. Phosphatidylcholine and sphingomyelin, for example, are inherently charged, efficiently ionized, and readily detected. Other lipids bear readily ionizable functional groups. Even neutral lipids may be nondestructively ionized by inducing them to form noncovalent gas-phase adducts with such carrier electrolytes as ammonia, alkali cations, formate or acetate.

Many simple lipids, including short chain fatty acids and sterols, are sufficiently volatile, either unmodified or after derivatization, that they can enter the gas phase in neutral form, whereupon they can undergo separation by gas chromatography (GC) before entry into the mass spectrometer. Complex lipids, covalent assemblies of simple lipid building blocks, often lack the volatility needed for gas chromatographic separation. Most complex membrane lipids fall into this category. About a decade ago such lipids were routinely ionized from dried residues by fast atom bombardment (FAB-MS) or a variety of other "soft ionization" methods then in use. The advent of matrix-assisted laser-desorption ionization (MALDI) and electrospray ionization (ESI) techniques have greatly expanded the range of options available for nondestructive ionization of complex lipids, and these ionization methods have been wedded to a variety of innovative instrument configurations, each of which offers specific advantages and disadvantages. This review focuses primarily on the analysis of complex lipids using ESI, atmospheric pressure chemical ionization (APCI), and MALDI techniques with mass analyzers of quadrupole and time-of-flight (TOF) design. New MS-based imaging techniques are also discussed.

MS SAMPLE INTRODUCTION, IONIZATION STRATEGIES, AND SCAN MODES

MS is a powerful technique that enables separation and characterization of lipid molecular species according to their mass-to-charge ratio (m/z). The combination of sensitivity, selectivity, speed and the capability of obtaining structural information for unknown lipids makes MS an ideal method for intact lipid molecular species analysis. Its essential components include a sample inlet, ion source, mass analyzer(s), detector, and data handling system.

Sample Introduction

Sample introduction depends first on the choice of ion source. ESI and APCI utilize samples in solution, while MALDI requires a crystalline sample. Liquid samples can be infused directly into the mass spectrometer, or indirectly *via* a liquid chromatography (LC) system. Whether to use direct infusion-MS or LC-MS depends on the complexity of the biological sample and the interest in detecting minor lipid species. MALDI samples can be prepared from the total extract, from LC fractions, or from tissue sections (for imaging).

Direct infusion, or analysis of each analyte in the presence of all the analytes, has the advantage of simplicity and application to a wide range of lipid species. Direct infusion of liquids is performed with a syringe pump or autosampler operating at fL/min flow rates. The composition of the sample entering the instrument is constant so ratios of internal standards to compounds of interest are constant throughout the analysis, making quantification straightforward. The main drawbacks of direct infusion are potential difficulties in resolving isobaric compounds, especially in the absence of tandem MS techniques, and a risk of ion suppression that may lead to decreased sensitivity, especially in the analysis of minor lipid species.

Utilizing LC-MS to separate lipids into either different lipid classes or molecular species within a class results in a less heterogeneous sample at each point of time during the analysis. Chromatographic resolution of a complex mixture into a series of temporally focused eluent peaks allows detection of less abundant lipid molecular species because there is less competition for the available ion current and because the concentrated peak is more likely to stand out against chemical background noise. LC-MS is particularly useful for upstream separation of isobaric lipids. The main drawbacks of LC-MS are more complicated method development, which includes optimization of LC separation conditions (mobile phase, pH, flow rate), lower throughput due to longer analysis times, and increased complexity of quantification due to separation of standards from analytes and differences in chromatographic peak shapes. When solvent gradients are used to control the LC separation, changes in solvent composition can alter the performance characteristics of the ion source, and thereby change detection sensitivity in a dynamic manner. Moreover, DeLong et al. (18) reported that a selective loss of particular phospholipid molecular species could occur during LC separation, rendering quantification inaccurate.

Ion Sources

After sample introduction, ionization technnology is necessary for transfer of analytes (lipid molecular species) from a liquid (ESI and APCI) or crystalline (MALDI) phase into a mass analyzer operating under vacuum. There are several types of ion sources for creating charged species. The choice of ionization technique depends on type of analyte, sample preparation, separation technique and compatibility with the available mass analyzer. Below we discuss the commonly used soft ionization techniques (ESI, APCI and MALDI) for complex lipid profiling.

ESI (Electrospray Ionization)

ESI/MS was first conceived in the 1960s by Malcolm Dole, but was developed for practical use by Fenn and co-workers (19, 20). ESI is a technique for production of gas-phase ions from molecules in a solution. The gas-phase forms are typically (intact) molecular (or "pseudomolecular") ions. For this reason, ESI is classified as a "soft" ionization technique.

Many theories of the physical processes involved in ion generation during ESI have been proposed and validated in some detail (20-23), but the mechanism is not fully understood. The principle of the ESI process is schematically represented in Figure 1. Initially, the LC effluent or direct infusate is pumped into the ESI needle through capillary tubing. The narrow orifice at the end of the ESI needle and the mechanical forces imparted on the solution passing from the orifice create a mist of small droplets in the ionization chamber. If an electric potential (approximately 2-6 kV) is applied between the end of the capillary tube and the open entry port into the mass analyzer, oxidation/reduction processes cause the droplets to carry a net charge. The polarity of this droplet charge drives ionization of target analytes during rapid desolvation processes. Gas phase ions are then harvested by acceleration toward the charged mass analyzer inlet port. The applied potential can be either positive or negative depending on the substance to be analyzed. As shown in Figure 1, if a positive electric potential is applied to the end of a capillary tube and a negative electric potential is present at the entrance

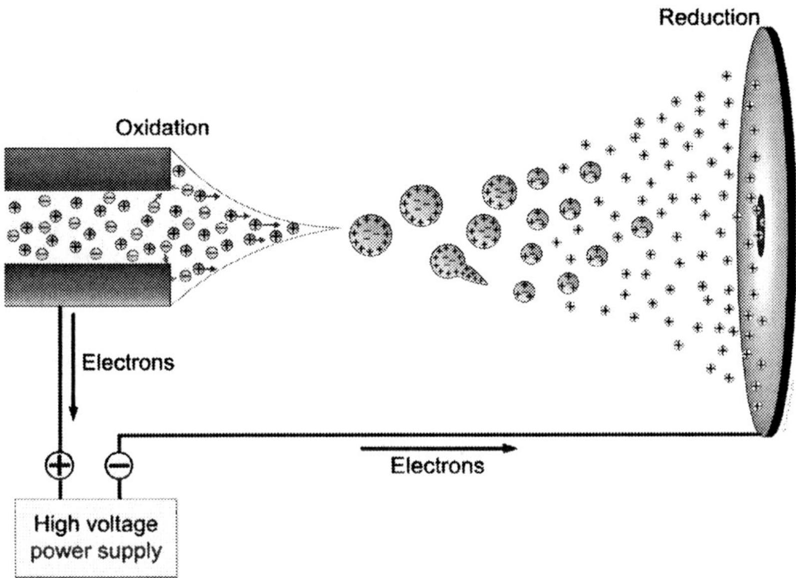

Figure 1. Electrospray ionization in positive ion mode.

of the mass analyzer in the positive-ion mode, droplets carry net positive charges. The resulting electric field causes charge separation of ions in the solution, with positive ions moving towards the liquid surface and negative ions migrating back into the liquid. The droplets are desolvated, by passage through a curtain gas. During desolvation, the coulombic force between ions is dramatically increased. Once this force exceeds the surface tension of the solvent, the droplets explode to form smaller droplets. This cycle is iteratively repeated at atmospheric pressure until molecular ions are generated prior to their entrance into the mass analyzer (21). The exact mechanism of ion formation, whether it is by ion evaporation (ion-evaporation model) or by complete solvent removal (charge-residue model) from the charged droplet, is debatable, and different mechanisms may apply in different situations (24). The process of ESI has been described in more detail elsewhere (21, 22, 25, 26).

Atmospheric Pressure Chemical Ionization (APCI)

Like ESI, APCI may be classified as a "soft" ionization technique, one which produces minimal fragmentation of the molecules being ionized. APCI works by evaporating solvent from an analyte solution sprayed into a concentric flow of heated sheath gas. Analyte molecules are ionized by chemical interactions with a gas-phase plasma created from the sheath gas *via* corona discharge from an electrode located some distance away. Electrolytes added to the analyte solution and the chosen composition of the sheath gas can be used to engineer the type of ionization events which occur.

MALDI (Matrix-Assisted Laser Desorption/Ionization)

MALDI is another soft ionization method. Rather than being presented as a solution, a sample is embedded in a chemical matrix that allows the production of intact gas-phase ions when a laser is aimed at the matrix. The sample is prepared by mixing an excess of matrix with the sample, or, for imaging, matrix is placed directly on thin tissue sections. A typical matrix is an aromatic acid with a chromophore that absorbs ultraviolet light of the wavelength produced by the laser (typically a N_2 laser), causing a small part of the target substrate to vaporize and ionize (27).

MS Scan Modes

A number of mass analyzers are available, including quadrupole, ion trap, TOF, ion cyclotron resonance and sector instruments; all separate charged species according to their *m/z* ratio. The most widely used instrument for lipidomics is the triple quadrupole, which is a tandem MS instrument. Some reasons for the popularity of this instrument are its large dynamic range (typically four to six orders of magnitude) and the ability to perform precursor and neutral loss scans. In a triple quadrupole instrument, the middle, field-free quadrupole focuses and transmits almost all ions and may be used as a collision cell for fragmentation reactions. The type of fragmentation occurring in quadrupole instruments is referred to as collision-induced dissociation (CID). The fragmentation is performed by collision of a selected ion with an inert gas or nitrogen. As a result of the collision, the internal energy of the ion may be increased by conversion of kinetic energy into internal energy (28, 29). The choice of collision gas, its pressure and the collision energy affect the degree of fragmentation. The fragment ions can be analyzed in the second mass analyzer (Q3). In addition to structural information, tandem MS provides higher sensitivity, specificity and reduced chemical background by selecting a molecular ion of interest from a number of ions present in the mass spectrum. Selection of scan modes (Table 1) specific for the different lipid classes, as described in the following sections, allows resolution of lipid molecular species in a complex mixture.

Table 1. Triple quadrupole mass spectrometer scan mode summary.

Scan	Q1	Q2	Q3	Record
Product	Fixed	On	Scanning	Detector vs. Q3 *m/z*
Precursor	Scanning	On	Fixed at characteristic (charged) fragment *m/z*	Detector vs. Q1 *m/z*
Neutral loss	Scanning	On	Scanning with m/z offset from Q1 of characteristic (neutral) fragment *m/z*	Detector vs. Q1 *m/z*

Product Ion Scan

Product ion scans are the most common type of scan performed by tandem MS instruments. In this scan mode, an ion of interest with a given *m/z* value is selected in the first mass analyzer. The selected ion is passed into the collision cell and fragmented by collision with inert gas or nitrogen. The product ions are then analyzed with the second mass analyzer. Fragment ions arising from the molecular ion are detected and recorded (28). The method is valuable for characterization of unknown structures and has also been used to confirm the identity of known lipid molecular species, such as glucosylceramides (30), sphingomyelins (31), phospholipids (32,33) and sulfatides (34). Product ion scans can be performed by any tandem MS instrument.

Precursor Ion Scan

A precursor ion scan is based on independent operation of two mass analyzers positioned inside a tandem MS on each side of the collision cell. A true precursor ion scan can be performed only by a triple quadrupole instrument, although product ion scan data from other types of instruments can be rearranged to simulate a precursor ion scan. In the precursor-ion scan mode, the first mass analyzer (Q1) of a triple quadrupole MS is scanning, and the second analyzer (Q3) is set to transmit a constant *m/z* ion. Ions from the first mass analyzer are recorded only when they produce, in the collision cell, a fragment of the specific *m/z* at which the second analyzer has been set. A precursor ion scan can selectively detect precursor ions corresponding to the molecular species in a lipid class, because those species produce the same characteristic head group-derived fragment ion upon CID (24, 28, 35). Likewise, a precursor ion scan can detect a group of complex lipids containing a common fatty acid, because those species can produce the same fatty acyl fragment upon CID (e.g., 36, 37).

Neutral Loss Scan

During a constant neutral loss scan, both Q1 and Q3 are scanned in parallel with a constant difference in *m/z* between the two analyzers. Like the precursor ion scan, a neutral loss scan is selective for a particular functional group, and also can be performed directly only by a triple quadrupole instrument. A neutral loss scan is used when the characteristic common fragment for a group of related compounds is uncharged. Like precursor ion scans, neutral loss scans have been applied to identify phospholipid molecular species within a class (6, 36, 38) and those containing common fatty acids (39, 40) from biological mixtures of lipids.

Characteristic Precursor and Neutral Loss Scans

Table 2 shows some now-standard and some recently described precursor and neutral loss scans utilizing characteristic fragments of specific polar lipid classes (32, 35, 37, 38, 41-50). Acyl-chain-specific precursor (36, 37) and acyl-chain-specific neutral loss (40) scans have also been described.

Table 2. Some precursor and neutral loss scan modes utilizing characteristic fragments generated by ESI for analysis of polar complex lipids as classes.

Lipids analyzed	Polarity	Ion analyzed	Scan mode	References
Phospholipids				
phosphatidylcholines	+	$[M+H]^+$	Precursors of *m/z* 184	38
	+	$[M+H]^+$	Neutral loss of 183	37
	–	$[M+Cl]^-$	Neutral loss of 50	37
phosphatidylethanol-amines	+	$[M+H]^+$	Neutral loss of 141	38
	+	$[M+H]^+$	Precursors of *m/z* 196	38
phosphatidylglycerols	–	$[M-H]^-$	Precursors of *m/z* 153	38
	–	$[M-H]^-$	Precursors of *m/z* 227	41
	+	$[M+NH_4]$	Neutral loss of 189	35
phosphatidic acids	–	$[M-H]^-$	Precursors of *m/z* 153	38
phosphatidylinositols	–	$[M-H]^-$	Precursors of *m/z* 241	32, 38
	+	$[M+NH_4]$	Neutral loss of 277	35
phosphatidylinositol phosphates	–	$[M-H]^-$	Precursors of *m/z* 321	42, 43
phosphatidylinositol bis-phosphates	–	$[M-H]^-$	Precursors of *m/z* 401	43
phosphatidylinositol tri-phosphates	–	$[M-H]^-$	Precursors of *m/z* 481	44
phosphatidylserines	–	$[M-H]^-$	Neutral loss of *m/z* 87	38
	+	$[M-H]^+$	Neutral loss of 196	38
Sphingolipids				
sphingomyelins	+	$[M+H]^+$	Precursors of *m/z* 184	38
ceramides (18C sphingosine)	+	$[M+H]^+$	Precursors of *m/z* 264	45
	–	$[M-H]^-$	Neutral loss of *m/z* 240 and 256	46
ceramides with 2-hydroxy fatty acids	–	$[M-H]^-$	Neutral loss of 240 and 327	46
sulfatides	–	$[M-H]^-$	Precursors of *m/z* 97	34
cerebrosides	+	$[M+H]^+$	Precursors of *m/z* 264	45
	–	$[M+Cl]^-$	Precursors of *m/z* 89 and 179	47
	+	$[M+Li]^+$	Neutral loss of 162 and 210	47
gangliosides	–	$[M-2H]^{2-}$	Precursors of *m/z* 290	48, 45
Other acyl lipids				
diacylglycero-trimethylhomo-serines	+	$[M+H]^+$	Precursors of *m/z* 236	Welti, previously unpub-lished.
sulfoquinovosyl diacylglycerols	–	$[M-H]^-$	Precursors of *m/z* 225	41, 49
monogalactosyl-diacylglycerols	+	$[M+Na]^+$	Precursors of *m/z* 243	41, 50
	+	$[M+NH_4]^+$	Neutral loss of 179	Esch & Welti, previously unpub-lished.
digalactosyl-diacylglycerols	+	$[M+Na]^+$	Precursors of *m/z* 243	41,50
	+	$[M+NH_4]^+$	Neutral loss of 341	Esch & Welti, previously unpub-lished.

ESI-MS AND ESI-MS/MS APPROACHES TO LIPIDOMIC ANALYSES

LC-MS

As mentioned previously, one distinction among lipidomic strategies is whether or not chromatography is utilized. Han and Gross estimate that they are able to detect and quantify lipid species accounting for over 90% of the mass of a biological (animal-derived) lipid mixture using a direct infusion ESI-MS and ESI-MS/MS method (40). The detection and quantification of particular minor lipid species, as well as the resolution of isobaric species, may, however, be aided by LC analysis. A comprehensive scheme for analysis of phospholipids, utilizing LC-MS/MS, has recently been described by Taguchi and co-workers (35). The authors utilize a combination of neutral loss and precursor scans to identify lipid molecular species, both in terms of lipid class and individual fatty acyl species. They state that use of LC allows them to identify some minor species which cannot be assigned unambiguously by analysis of data obtained from direct infusion. Although these workers indicate that quantification can be performed, to our knowledge no information about the quantitative aspects of the analysis has yet been provided. On the other hand, there are many examples of the use of LC-MS or LC-MS/MS to identify and quantify smaller groups of lipid species. For example, Merrill and co-workers demonstrate the utility of normal- and reverse-phase LC, in conjunction with ESI-MS/MS detection and identification, to separate isometric and isobaric species, such as glucosylceramides and galactosylceramides (summarized in 45). Isaac and co-workers utilized LC-MS to separate phosphatidylcholine and sphingomyelin molecular species, pushing the limit of detection down to the low femtomole range (51).

ESI-Direct Infusion Strategies

There are several direct infusion strategies for identification and quantification of lipid species. Recently, these strategies have been described and evaluated in an excellent review by Han and Gross (40), and considerations related to quantification, including choice of internal standards, correction of spectral data for isotopic variants of each molecular species, and comparison of the sample peaks with those of the internal standards have been discussed. Quantitative analytical strategies include analysis of single MS spectra (MS1) followed by product ion scanning for species identification (e.g., 52, 53), utilization of intrasource separation or collection of multiple MS1 spectra under varied ionization conditions, before proceeding with tandem mass spectral analysis (e.g., 11), collection of a series of precursor and neutral loss scans characteristic of each lipid class, sometimes followed by product ion analysis for identification of individual acyl species (e.g., 1, 6, 38), and analysis that combines acyl chain precursor or neutral loss scanning with one of the methods for detection of species within a head group class (e.g., 12, 36, 37, 39, 40). A comprehensive analysis by any of these methods involves collecting a series of spectra from each sample and analyzing and combining the spectral data. Some studies have identified molecular species only in terms of total acyl carbons and double bonds (e.g., 38, 54), while others have identified each lipid at the true molecular species level (e.g., 36, 37).

Challenges for Comprehensive ESI-MS Profiling Strategies

The effectiveness of electrospray-based lipidomic strategies for comprehensive and high-throughput analysis depends on several factors: (1) sensitivity and selectivity of the MS scanning methods, (2) availability of appropriate and quantified internal standards, and (3) availability of a method to process the mass spectrometric data rapidly. In the past few years, many new MS scanning methods have been described, so that this factor is probably the least problematic at the present time. Depending on the suite of MS methods chosen, from several to many internal standards are required. In general, methods that involve collection of MS1 spectra require fewer internal standards than those methods that depend only on precursor and neutral loss scanning, because standards of one class may be used to quantify compounds in another class and a single standard for each class can be sufficient. On the other hand, precursor and neutral loss scanning methods generally require two internal standards of each head group class (38). Standard compounds are ideally non-naturally occurring lipids or isotopic variants of the same head group class, with masses outside the range of the naturally-occurring compounds. Accurate quantification of standards requires analysis of phosphate or fatty acid content. In some cases, there are no appropriate compounds commercially available; thus labs are required to produce their own standards by hydrogenation or semi-synthetically. Commercial availability of appropriate and quantified mixtures of standards would not only make profiling easier, but would eliminate one factor (variability in quantification of standards) that potentially leads to variability in quantification among practitioners.

Wide access to standardized methods for data processing would also pave the road to more uniform and comparable lipid profiles. Some of the strategies currently used are discussed near the end of this review under the heading, "INFORMATIC CHALLENGES, Data Processing". Perhaps before use of uniform data processing methods is practicable, there will have to be some consensus as to what constitutes the "best" MS strategy, i.e., a strategy that provides an appropriate balance between comprehensive and rapid analyses and which is simple enough to be performed in a reproducible manner in different labs. In our opinions, the current front-runner strategies are direct-infusion strategies that utilize precursor and neutral loss scans.

Coordination Ion Spray-Mass Spectrometry (CIS-MS) can Enhance ESI of Non-Polar Lipids

CIS-MS is an ESI-based technique that utilizes metal ion adduct formation to ionize and detect non-polar compounds that are difficult to ionize (55). A metal salt is added to the mobile phase (direct infusion or chromatography) or by post-column addition, and the solution is ionized in positive ion mode. Silver is usually used as the metal, but other metals such as copper, nickel, platinum, palladium (transition metals) may be used to form highly stable pi or pi-allyl complexes with unsaturated compounds (56). Since silver has nearly equal amounts of the 107 and 109 amu isotopes, the adduct peaks in the mass spectrum are distinctive and easy to recognize. Moreover, the compounds are subject to less thermal stress than when using APCI (next section), and the sensitivity is significantly higher (55). However, a major drawback of this method is that Ag_2O is deposited on the MS ion optics, requiring regular cleaning.

Bayer et al. (55) detected a wide selection of compounds such as olefins, polyolefins, and aromatic compounds, including steroids, vitamins of the D and E families, carotenoids, polystyrenes, terpenes, and unsaturated fatty acids. A number of recent articles report success with coordination ion spray of molecules using ESI-MS, either by direct infusion or using a chromatography step. Compounds analyzed include endocannabinoids (56), capsaicinoids (57), intact high-molecular wax esters in jojoba oil (58), vitamin E derivatives and homologs (59, 60), non-polar compound classes in plant extracts (61), triacylglycerols in vegetable oils (62), lipid peroxidation products (63-67), and ginsenosides (68). Ho and co-workers (69) used product ion analysis of cobalt (II) ion complexes of phosphatidylethanolamines, phosphatidylglycerols, and phosphatidylserines to identify not only the molecular species, but to obtain regiospecific information about the two fatty acyl substituents.

APCI-MS FOR LIPIDOMICS

Atmospheric Pressure Chemical Ionization

ESI-MS/MS instruments often have an APCI source available as an accessory. Several instrument manufacturers have begun offering dual APCI-ESI sources which allow rapid switching between ionization modes. Some lipid analytes, particularly those that are less polar, ionize poorly by ESI, but better *via* APCI. Arguably, steroids tend to fall into this category (27, 70-79). Concurrent measurement of sterols as part of the broader lipid profile is one example where a dual APCI-ESI source may be of particular value in a lipidomics context. Even in cases where ESI and APCI deliver comparable ionization efficiency, significant differences in the way lipids respond to these two ionization methods can be exploited as complementary detection options.

The versatility of an APCI-MS approach was summarized recently by William Craig Byrdwell (80), an innovative pioneer in successful application of this technique to lipid analysis. His review detailed the work culminating in an APCI-MS method for accurate quantification of triacylglycerol homologs in plant oils (81, 82). The suite of fatty acids appearing in triacylglycerols may adopt a daunting array of permutations, because any three may take several different relative positions on the glycerol backbone to form positional isomers. As a class of inherently neutral molecular structures, triacylglycerols are difficult to ionize efficiently, although suitable APCI (and also ESI) conditions can produce almost exclusively an intact protonated molecular ion, or foster adduct formation with a 'counterion' such as Li^+, Na^+, K^+, Cs^+ (83 and references therein). Ammonium adducts have proven particularly useful for APCI analysis (83, 84). During the APCI process, some triacylglycerol species (notably those with multiple unsaturations) do tend to remain intact, and appear in MS spectra as $[M+H]^+$ or as positively-charged adducts. Other triacylglycerol species (i.e., those with few or no unsaturations) tend to lose an acyl chain, thereby forming diacylglycerols. Some intermediately unsaturated triacylglycerol species appear as both forms in the same spectrum. Acyl chains bearing hydroxyl groups tend to undergo a neutral water loss in the APCI source. Byrdwell and co-workers (principally, W. E. Neff) developed a detailed knowledge of these characteristics to help identify and hence quantify the amount of each successive triacylglycerol analyte present in the effluent stream of a reverse-phase column. Chromatographic separation further differentiated the triacylglycerol analytes on the basis of their acyl chain characteristics (82), and reliance upon chromatographic separation has persisted (85-87). As noted by Byrdwell and Neff (83), and as described previously, others have successfully used ESI-MS/MS for triacylglycerol analysis (39, 88, 89) and ESI-MS/MS does offer greater sensitivity than APCI. Byrdwell and Neff themselves recently revisited ESI as a method for triacylglycerol ionization (84).

MALDI-MS FOR LIPIDOMICS

MALDI technologies provide another attractive alternative to ESI-triple quadrupole MS for a number of challenging qualitative lipid analytical tasks (90-98). A recent review by Schiller et al. (90) provides a lucid description of MALDI-MS, its use in profiling individual lipid classes, and its general applicability to a variety of lipid sources. One of the major advantages of a MALDI-MS, and particularly MALDI-TOF-MS, is its rapidity, with analysis of individual samples often taking less than one minute (90). For this reason, the consortium of lipidomics groups that are carrying out the lipid identification and quantification project called "LIPID MAPS" plans to eventually transfer the analytical methods developed with ESI-MS to MALDI-MS (99).

When a MALDI source is used in conjunction with a TOF analyzer, the intensities of intact ions produced by the MALDI source can be measured directly in a simple MS spectrum. Schiller et al. (90) identify several MALDI-TOF methods that can be applied to lipid analysis. More recently, Jackson et al. used MALDI-MS to identify molecular ions for a variety of lipid classes in the lipidome of rat cerebellum. Cholesterol, phosphatidylcholines, sphingomyelins, phosphatidylethanolamines, phosphatidylinos-itols, sulfatides and gangliosides were all detected in the MS spectra obtained (100).

There are problematic drawbacks, however, to MALDI-TOF technology. Complete MALDI-TOF experiments may be carried out in seconds on a high throughput basis, but careful sample preparation and information about lipid class-specific instrument response characteristics is crucial to obtaining consistent reproducible experimental results. The low-mass region of MALDI-TOF spectra tends to be dominated by large peaks representing matrix ions and clusters of matrix ions. These can obscure or overlap the molecular ions of smaller lipids. As is true for other ionization techniques, various lipid classes have different ionization, transmission and detection efficiencies. For any given lipid class, detection sensitivity tends to decrease as a function of increasing lipid mass (i.e. acyl chain length) with the double-bond content having a secondary influence in

MALDI ionization efficiency (100; see also 101 and 102, as cited in 100). Limits of quantification and limits of detection for different lipid species can range from upwards of 1 nanomole per spot in the case of phosphatidylinositol triphosphate (103) down to 50 femtomoles for triacylglycerols (104). These limits depend on the chemistry of the lipid in question, the choice of matrix, and the manipulative expertise.

MALDI ion sources have been interfaced to TOF, TOF-TOF, ion cyclotron resonance and triple quadrupole analyzers. Compared to a quadrupole mass analyzer, the TOF mass analyzer has a very limited linear dynamic range typically two to three orders of magnitude. Ion cyclotron resonance analyzers have much higher resolution capabilities, but they also suffer from a limited dynamic range compared to triple quadrupole instruments. Recently, quantitative small molecule characterization was achieved by linking a MALDI source to a triple quadrupole mass analyzer (105). In another intriguing development, MALDI-ion mobility-TOF-MS was used to separate isobaric overlaps between lipids and non-lipid analytes (peptides or nucleotides) present in the same sample (106). Instruments of the MALDI-TOF/TOF or -quadrupole/TOF configuration are often chosen to obtain product ion spectra. When an instrument of quadrupole/TOF design was equipped with a MALDI source, product ion spectra of glycosphingolipids were found to be "almost identical to those obtained by electrospray", "with no loss of sensitivity with increasing mass as was observed from the corresponding ions produced by electrospray" (107; see also 101 and 102, as cited in 90).

MALDI-MS spectra obtained from complex total lipid extracts can become quite crowded with peaks. Ion suppression contributed by the more abundant or readily detectable lipid classes may obscure less abundant or less efficiently detected lipids. Accordingly, the more informative MALDI analyses are often conducted upon lipid preparations that have been pre-fractionated by some suitable separation technology such as thin-layer chromatography or LC. The logistical requirements of such fractionation or sample preparation methods must be taken into account when planning to harvest the "high-throughput" capabilities offered by MALDI-MS.

MALDI AND TOF-SECONDARY ION MASS SPECTROMETRY (TOF-SIMS) FOR LIPID IMAGING

MS tools are now being utilized to investigate the physical distribution of lipids in cells and tissues without the need for separation of these compartments and extraction of the lipids. The main advantage of MS imaging tools over "traditional" microscopic methods is their capacity to localize numerous known and unknown compounds simultaneously in cells and tissues. Using MS tools, the target compounds don't need to be identified *a priori*, and unknown compounds can be retrospectively identified.

Two major techniques have been developed to directly probe the distribution of biomolecules in living tissues and cells: MALDI coupled to a TOF-MS and TOF-secondary ion mass spectrometry (SIMS). Using MALDI, Caprioli and co-workers have directly analyzed large biomolecules, mainly proteins and peptides, in tissues (see reviews: 108-117). TOF-SIMS has also been used to localize molecular and elemental ions (see reviews: 113,117-123). Only recently, have lipids been specifically targeted for direct analysis using MS imaging tools.

MALDI Imaging

MALDI is well-suited for the direct analysis of biomolecules in tissues because of its high sensitivity, high tolerance for salts and other contaminants, and ability to produce ions of a large mass range with little fragmentation. One reason why little lipid imaging work using MALDI has been performed is that most lipids have masses below 1000 amu, and this spectral region is complicated by chemical noise and interference from matrix ions. Imaging lipid distribution in tissues (and cells) by MALDI does not require special techniques apart from the ones used for proteins and peptides. However, special care should be taken in the preparation of the biological materials to conserve the chemical integrity of the lipids. Table 3 presents papers published on lipid distribution using MALDI and a related laser ionization/desorption method (LDI) (100, 124-129).

To perform MALDI imaging, fresh-frozen tissues (100, 124, 125, 127-129) or fixed tissues (124) are cut in thin sections and the matrix is deposited directly on the tissue sections (see Table 3 for the matrix used in the various studies). Sluszny et al. (126) used colloidal silver deposited on the plant tissues to increase ionization (cationization) of cuticular waxes. The greatest difficulties in successful MALDI image analysis are in obtaining uniform deposition of organic MALDI matrix (and colloidal silver) and homogeneous matrix crystallization without solvent extraction effects during drying. Tempez et al. (129) demonstrated that implantation of gold clusters in tissue sections, desorption/ionization of lipids using a laser microprobe, and detection by ion-mobility MS may offer significant advantages as an alternative method, because these methods result in homogeneous, nondestructive and uniform matrix incorporation into the near-surface region of the biological tissue.

Only relative quantification has been attained thus far in MALDI imaging analyses. Rujoi et al. (124) note that many factors make quantification difficult, including differences in ionization efficiency among the different lipids and high variability in results. Many studies performed thus far, utilizing MALDI on thin sections, have only attempted to identify lipid metabolites, rather than to create a "map" of the section with lipid species densities throughout the section. Indeed, current MALDI imaging analyses provide spatial resolution of only 100 to 300 μm; this resolution is limited by the degree of homogeneity of matrix distribution in conjunction with the thickness of the tissue slices (124,130). Thus, MALDI imaging at this time is generally limited to discerning lipid compositional differences across tissues, rather than cells.

TOF-SIMS Imaging

TOF-SIMS (ion-beam-induced desorption) is a powerful method for imaging lipid distribution in biological samples with resolution of less than 0.5 to 5 μm (subcellular resolution) (e.g,, 130, 131-134). A sample/surface is bombarded with a beam of energetic ions (termed the `primary ions'), and the ions desorbed

Table 3. Examples of laser desorption/ionization imaging studies that investigate lipid distribution in various biological materials.

Biological material	MS imaging method (matrix)	Lipid species studied	Reference
Bovine lens	MALDI-TOF-MS (p-nitroaniline)	Phosphatidylcholines Sphingomyelins	124
Muscles of legs of the mouse model with Duchenne muscular dystrophy	MALDI-TOF-MS (a-cyano-4-hydroxy cinnamic acid)	Phosphatidylcholines	125
Arabidopsis thaliana and maize leaves	LDI-TOF-MS (with colloidal silver deposition)	Cuticular waxes (alkanes, alcohols, aldehydes, ketones, fatty acids, from C26 to C32; esters from C42 to C48)	126
Rat brain	MALDI-TOF-MS/TOF-MS (2, 6-dihydroxyacetophenone; 2, 6-dihydroxyacetophenone /lithium chloride)	Phosphatidylcholines	127
Rat brain	MALDI-TOF-MS (2,6-dihydroxyacetophenone; 2, 5-dihydroxybenzoic acid)	Phosphatidylcholines, phosphatidylethanolamines, phosphatidylethanolamines, sphingomyelin, phosphatidylinositols, sulfatides, gangliosides	100
Rat brain	MALDI-TOF-MS MALDI-Ion Mobility-TOF-MS (6-aza-2-thiothymine; 2, 6-dihydroxyacetophenone)	Phosphatidylcholines, phosphatidylethanolamines, sphingomyelins	128
Rat brain	Matrix Implanted Laser Desorption Ionization (MILDI)-Ion Mobility-TOF-MS (using $Au_{400}{}^{4+}$ cluster implantation procedure)	Phosphatidic acids, phosphatidylethanolamine, phosphatidylcholine, cerebrosides, phosphatidylserine, sphingomyelins	129

('secondary ions') from the sample/surface are captured and analyzed by a TOF-MS (117, 119, 120, 123). SIMS practitioners tend to refer to the element or compound producing the ions as the "ion source". TOF-SIMS may be conducted in the dynamic or the static mode. The dynamic mode occurs when the total primary ion density applied exceeds $\sim 1\times 10^{13}$ primary ions/cm^2 during analysis. This mode is used for elemental analysis and depth profiling. The static mode uses less than $\sim 1\times 10^{13}$ primary ions/cm^2. This mode is used for surface profiling (1-2 molecular layers) and the analysis of molecular ions (117, 123). The static mode is used for profiling lipids from biological materials.

Among the many factors to consider before analyzing a tissue/cell with TOF-SIMS, sample preparation and choice of primary ion source are particularly important. These depend on the analytical requirements. Conventional primary ion sources (Ga^+, In^+, ReO_4^-, Cs^+) have proved useful for visualization of some lipid and lipid-derived ions, including phosphocholine, a characteristic fragment of phosphatidylcholine species (135-137), 2-aminoethylphosphonolipid, the phosphonolipid analog of phosphatidylethanolamine (137), cholesterol (see references above; 135), low molecular weight lipids such as fatty acids (138, 139), alkanes in plant waxes (140) and diterpene phenol (141). However, extensive fragmentation of organic molecules and insufficient yields of molecular ions has limited characterization of biological materials by TOF-SIMS using conventional ion sources (134, 142). New primary ion sources such as gold (Au_n^+), bismuth (Bi_n^+) and buckminsterfullerene (C_{60}^+) have recently been developed to overcome these limitations (120) and to produce spectra richer in large fragments characteristic of complex lipids.

The use of polyatomic (ion clusters) guns such as gold (Au_n^+), bismuth (Bi_n^+) and buckminsterfullerene (C_{60}^+) allows the secondary ion yield of large organic molecules to increase by several orders of magnitude, even while maintaining submicron to micron lateral resolution. Using lipid standards and/or directly *in situ*, it was demonstrated that these ion sources have great potential for profiling simple and complex lipids as well as polar and non-polar lipids. For example, phosphatidylcholines, phosphatidic acids, phosphoinositides, phosphatidylglycerols, phosphatidylserines, sphingomyelins, sulfatides, galactosylceramide, cholesterol, antioxidant compounds (vitamin E, coenzyme Q9), fatty acids, diacylglycerols and triacylglycerols have been detected using new primary ion chemistries (125, 130, 132, 134, 142, 143). Table 4 summarizes examples of static SIMS imaging of lipid distribution in various biological materials.

Readers are referred to the papers cited in Table 4 (130-141, 144) and to reviews (113, 117-123) for more details about the preparation of sample. In general, the sample must be smooth and flat to limit ionization and mass measurement defects. The TOF-SIMS studies in which more lipid species were identified used a sample strategy of freezing the fresh tissue, cutting in thin slices and freeze-drying under vacuum (130, 134).

Promising New Approaches for "Imaging" Biological Materials

Besides TOF-SIMS and MALDI, other approaches have been used to desorb compounds from surfaces, and it is likely that the toolbox for imaging will

Table 4. Examples of static SIMS imaging of lipid distribution in various biological materials.

Biological material	MS imaging method (beam type)	Lipid species studied	Reference
Intact microbial cells (bacteria)	Static SIMS (TQ-MS) (ReO_4^-)	Fatty acids	138
Rat retina	Static SIMS (TOF-MS) (Ga^+)	Fatty acids	139
Human leucocytes Rat kidneys	Static SIMS (TOF-MS) (Ga^+) (with silver deposition)	Cholesterol, phosphocholine	131,144
Rat cerebellar cortex	Static SIMS (TOF-MS) (Bi_{1-7}^+)	Phosphatidylcholine and sphingomyelin (phosphocholine), cholesterol, galactosylceramide	132
Muscles of legs of the mouse model with Duchenne muscular dystrophy	Static SIMS (TOF-MS) (Au_3^+cluster)	Phosphatidylcholines (phosphocholine), antioxidant compounds (vitamin E, coenzyme Q9), phosphatidic acids, phosphoinositides, fatty acids, diacylglycerols, triacylglycerols, cholesterol	130
Prunus laurocerasus leaf	Static SIMS (TOF-MS) (Ga^+, Cs^+)	Epicuticular waxes (alkanes C29 and C31, family of acetates with even-number hydrocarbon chains in the C20-C34 range)	140
Cryptomeria japonica heartwood	Static SIMS (TOF-MS) (Ga^+)	Ferruginol (diterpene phenol)	141
Human blood cells	Static SIMS (TOF-MS) (Ga^+) (with silver imprinting)	Cholesterol, phosphatidylcholine (phosphocholine)	133
Tetrahymena thermophila	Static SIMS (TOF-MS) (In^+)	Phosphatidylcholine (phosphocholine) 2-aminoethylphosphonolipid (phosphonolipid analog of phosphatidylethanolamine)	137
Mouse brain	Static SIMS (TOF-MS) (Au_3^+ cluster)	Cholesterol, sulfatides, phosphoinositides, phosphatidylcholines (phosphocholine)	134
Rat pheochromocytoma (PC12)	Static SIMS (TOF-MS) (Ga^+)	Phosphocholine, Cholesterol Hydrocarbon	135,136

continue to expand in the near future. Rubakhin et al. (117) suggest that a promising advance will be to use ESI in imaging, since the groups of Van Berkel (Oak Ridge National Laboratory) and Cooks (Purdue University) have demonstrated the direct sampling of surfaces and ionization of analytes using ESI (145-153). Finally, it was also demonstrated that APCI (154,155), as well as a new ionization source, called DART for "Direct Analysis in Real Time", using electronic or vibronic excited-state species (156), were able to analyze molecules directly on surfaces. In addition, Cooks and co-workers introduced a new method called desorption electrospray ionization (DESI) that allows the ionization of polar and non-polar compounds present on metal, polymer, mineral and biological surfaces (147-149, 157). Wiseman et al. (157) demonstrated the capacity of DESI for profiling phosphatidylcholine and sphingomyelin species under ambient conditions in biological tissues and the potential use of this approach for molecular imaging lipid distribution. The new techniques are promising, but their capacity to generate molecular maps, especially using lipids as targets, of cells and tissues at the high level of resolution offered by the TOF-SIMS, remains to be demonstrated.

SPECIALIZED DERIVATIZATION TECHNIQUES THAT CAN AID LIPIDOMIC ANALYSES

Analysis of complex lipids by soft-ionization techniques such as ESI, APCI, atmospheric pressure photoionization (APPI) and MALDI is generally done directly without modification or transformation of the molecules. While the development of APCI and APPI may aid in analysis of non-polar lipid species in particular, the detection and identification of specific lipid classes with defined functional groups (for example, sugar and amino groups) may be facilitated by derivatization. Derivatization strategies have been reviewed (158-166), including their application to detection of fatty acids, phospholipids and steroids (167-170). The main goals of derivatization (as noted by Halket and Zaikin (159), Halket et al. (162)), are increasing volatility, increasing thermal stability and catalytic stability, improving of chromatographic properties for investigation by GC-MS and LC-MS, obtaining new and additional structural information, studying the mechanisms of dissociative ionization, increasing sensitivity and selectivity in trace analysis and quantitative determination by selected-ion monitoring (SIM), increasing the ion yields and finally, chiral analysis. Derivatization can also be used simply to create mass shifts of specific lipid groups in complex mixtures in order to better resolve isobaric species (171).

Three examples illustrate potential uses for derivatization in lipidomic studies. A simple and fast procedure was used for derivatization of phosphatidylethanolamine and lysophosphatidylethanolamine molecular species directly from lipid extracts of biological samples. The lipid extracts were briefly treated with fluorenylmethoxycarbonyl (Fmoc) chloride; phosphatidylethanolamine and lysophosphatidylethanolamine species were selectively derivatized to their corresponding carbamates. The reaction mixture was directly analyzed by direct infusion ESI-MS in the negative mode or as a neutral loss of the Fmoc moiety. This procedure dramatically enhanced the analytical sensitivity and allowed identification and quantitation of low-abundance molecular species with a detection limit of attomoles per microliter for phosphatidylethanolamine and lysophosphatidylethanolamine analysis with a 15,000-fold dynamic range. By using this approach, zwitterionic phosphoethanolamine-containing species were converted to "mass-shifted" derivatized anionic species, rendering possible the resolution of isobaric lipid species (171). In the second example, sphingosine-1-phosphate and dihydrosphingosine-1-phosphate were directly acetylated in a crude lipid extract prior to LC-MS/MS. This derivatization modified the zwitterionic nature of sphingosine-1-phosphate and dihydrosphingosine-1-phosphate and made them analyzable by negative ion ESI LC–MS/MS. The resulting method was highly selective and sensitive, capable of reliable detection of less than 50 µmol of sphingosine-1-phosphate and dihydrosphingosine-1-phosphate derviatives (172). Finally, bioactive lipids (fatty acids, hydroxy fatty acids, prostaglandins and iso-prostaglandins) were derivatized with an electron-capturing group such as a pentafluorobenzyl moiety before LC analysis. The corona discharge used to generate ions under conventional APCI conditions provided a rich source of gas-phase electrons, so that suitable analytes underwent electron capture in the gas phase in a similar manner to that observed for electron capture negative chemical ionization in GC-MS studies. Enantiomers and regioisomers of diverse bioactive lipids were resolved and quantified with great sensitivity using normal-phase chiral chromatography and electron capture APCI-MS/MS coupled to a stable isotope dilution methodology (173). These examples demonstrate the utility of derivatization in special circumstances in lipidomics studies to resolve classes of lipids in complex mixtures, as well as to aid in the separation, quantification and structural identification of complex lipids.

INFORMATIC CHALLENGES

Three informatic challenges exist in MS-based lipidomics: (1) data processing, i.e., deriving quantitative information from mass spectra; (2) organization of data and derived information; (3) utilizing the derived information in biological contexts.

Data Processing

Mass spectral data on lipids have been reported in the literature in varied formats, from peak sizes associated with unassigned *m/z* values, to peak sizes of assigned lipid species, to lipid quantities calculated in comparison to internal standards. In general, individual labs have generated their own programs or templates for sorting mass spectral data and doing calculations. Details of calculational steps to obtain lipid quantities from triple quadrupole data are detailed by Han and Gross (40, 174, 175) and by Brügger et al. (38), and additional calculational methodology for combined phosphatidylcholine and sphingomyelin analysis was described by Liebisch et al. (176). Welti and co-workers (1) simply utilize Excel templates that contain information about the lipid targets, and formulas that sort and calculate the mass spectral data, and other labs have developed similar "in lab" methods. Recently, Taguchi et al. (35) described the establishment of a tool (http://metabo.umin.jp) called "lipid search" for identifying lipid species, based on mass spectral data. Further developments of this or similar systems, inclusion of quantification methods in the data processing tools, and dissemination of the tools are sorely needed. Jones et al. (95-96) detail a method to extract lipid profile data from MALDI-FTMS spectra by "binning" multiple ion, derived from each chemical species and detected at accurate *m/z* value.

Organization of Data and Derived Information: Databases

Lipidomics technologies generate large amounts of data, and a database or databases to accommodate the data and to provide access to data in a format that will allow utilization and integration with gene expression, proteomic, biochemical and physiological data are clearly needed. Currently, information is available on the web in several locations, and additional databases are under construction. The LIPID MAPS consortium is developing a database that contains both information about individual lipid species and data derived from biological experiments at http://www.lipidmaps.org/data/index.html. The Lipid Library, produced by W.W. Christie, contains a large variety of information about mass spectral analysis of lipids, including annotated mass spectra of some lipids and a comprehensive list of related literature, available at http://www.lipidlibrary.co.uk/. METLIN, http://metlin.scripps.edu/, a collaborative effort between the group of Ruben Abagyan and the Center for Mass Spectrometry at the Scripps Research Institute, is an easily searchable database containing mass spectral data on metabolites, but currently (December 2005), there do not appear to be data on complex lipids in the database. The Metabolite Database of the Human Metabolome Project, http://www.hmdb.ca/, at the University of Alberta, also is easily accessible, but currently contains information on only a few complex lipid species. At http://csbdb.mpimp-golm.mpg.de/, the Comprehensive Systems Biology Database (177) hosted by the Max Planck Institute of Molecular Plant Physiology has a database called MetabolomeDB that contains plant metabolite data, but again does not seem to currently include complex lipid data. The Comprehensive Systems Biology Database contains gene expression and biological data, as well as metabolite profile data. Pedro Mendes and co-workers at Virginia Polytechnic Institute are also constructing, for metabolite profiling data on the plant, *Medicago truncatula*, a relational database called DOME, which will contain information about mRNA levels and proteomic data, as well as metabolites. L.J. Wang and co-workers at Kansas State University (http://bioinformatics.k-state.edu/) are constructing a database, called LipidomeDB, primarily for complex lipid profiles.

Utilizing the Derived Information in Biological Contexts

No single approach to statistical analysis of lipidomic data has yet emerged. There is even no general agreement about the stage at which to apply statistical analysis—on raw spectral data or on quantified lipid species data. Workers in the Alex Brown laboratory (178, 179) detail a method of statistically analyzing mass spectra for differences among samples. The method involves data normalization and use of Shewhart control charts to compare spectral values with mean values from control samples. The technique is designed to facilitate identification, among large numbers of spectra from biological samples, of differences in sizes of mass spectral peaks, which are then translated into identified lipid species. The output of the technique is a "lipid array" that indicates the lipid species alterations.

Other workers have used the multivariate statistical method, principal component analysis (PCA), to classify samples and/or to determine which lipids are most strongly correlated with the sample identifications. Fang et al. (180) utilized PCA to identify lipid differences among bacterial species. Sutphen and coworkers (15) identified several lysolipids in plasma as biomarkers with potential to distinguish subjects with ovarian cancer from control subjects. Wang et al. (13) identified phosphatidylethanolamine and lysophosphatidylcholine species differences between Type 2 diabetic and control subjects using PCA of plasma phospholipids analyzed by LC-MS. Mortuza et al. (181) also used PCA on ESI-MS data on rats with phospholipidosis to identify a potential biomarker, lyso-bisphosphatidic acid. Davidov and co-workers (182) have performed multivariate statistical analyses, combining protein and metabolite analysis in wild-type and transgenic mice. They identified triacylglycerol and lysophosphatidylcholine levels as characteristic of the genotypes and have developed a network model to depict lipid metabolism alterations in the transgenic mice.

Although such multivariate statistical approaches have brought some success in identifying lipid species accounting for biologically significant changes, pattern recognition would likely aid in analysis of differences among lipid profiles. Alterations in levels of polar lipid molecular species are often subtle as compared to alterations in the levels of many other metabolites, and it is often difficult to establish the statistical significance of subtle change. On the other hand, multiple lipid molecular species are often affected by alteration of a single enzyme activity. This is because many lipid metabolic enzymes act on groups of substrates rather than single molecular species, and a particular polar lipid molecular species are tied metabolically to those species with the same or related head groups and to those species with the same or related acyl chains. A subtle change in the availability of one building block of a complex lipid, will affect a group of molecular species. Concerted small changes in a subgroup of lipid metabolites should be easier to detect in lipid profile data than changes in single metabolites, and it should be easier to establish the statistical significance of an overall change in the pattern of lipid species. Thus, more work needs to be done to optimally tailor available statistical approaches to analysis of lipid profiles.

OUTLOOK

Lipidomics is entering an exciting stage. Significant challenges remain in developing efficient strategies for data storage, analysis and mining. Developing techniques, including mass spectral imaging, have strong potential. Although no "best" analytical strategy has yet emerged, techniques are advanced enough that these have capability to generate data that will help us understand the function of genes and enzymes involved in lipid metabolism and lipid signaling, to understand physiological and cellular processes involving lipids, and to develop tools that will help diagnose and treat human, plant and animal diseases.

ACKNOWLEDGMENTS

The authors appreciate the helpful suggestions of Dr. Todd D. Willliams. We would like to thank Dr. Andreas Dahlin for drawing Figure 1. Work and instrument acquisition in the authors' laboratories is supported by grants from National Science Foundation (MCB 0416839 and DBI 0520140). Support of the Kansas Lipidomics Research Center is from National Science Foundation's EPSCoR program, under grant EPS-0236913 with matching support from the State of Kansas through Kansas Technology Enterprise Corporation and Kansas State University, as well as from Core Facility Support from NIH grant P20 RR016475 from the INBRE program of the National Center for Research Resources. This is contribution 06-159-B from the Kansas Agricultural Experiment Station.

REFERENCES

1 Welti, R., Li, W., Li, M., Sang, Y., Biesiada, H., Zhou, H.-E., Rajashekar, C.B., Williams, T.D. and Wang, X. (2002) J. Biol. Chem. 277, 31994-32002.

2 Nandi, A., Krothapalli, K., Buseman, C.M., Li, M., Welti, R., Enyedi, A. and Shah, J. (2003) Plant Cell. 15, 2383-2398.

3 Nandi, A., Welti, R. and Shah (2004) Plant Cell 16, 465-477.

4 Li, W., Li, M., Zhang, W., Welti, R. and Wang, X. (2004) Nat. Biotechnol. 22, 427-433.

5 Yan, W., Jenkins, C.M., Han, X., Mancuso, D.J., Sims, H.F., Yang, K. and Gross, R.W. (2005) J. Biol. Chem. 280, 26669-26679.

6 Schneiter, R., Brügger, B., Sandhoff, R., Zellnig, G., Leber, A., Lampl, M., Athenstaedt, K., Hrastnik, C., Eder, S., Daum, G.,Paltauf, F., Wieland, F.T. and Kohlwein, S.D (1999) J. Cell Biol. 146, 741-754.

7 Kim, H.Y., Bigelow, J. and Kevala, J.H. (2004) Biochemistry 43, 1030-1036.

8 Osawa, Y., Uchinami, H., Bielawski, J., Schwabe, R.F., Hannun, Y.A. and Brenner, D.A. (2005) J. Biol. Chem. 280, 27879-27887.

9 Masuda, S., Murakami, M., Takanezawa, Y., Aoki, J., Arai, H., Ishikawa, Y., Ishii, T., Arioka, M. and Kudo, I. (2005) J. Biol. Chem. 280, 23203-23214.

10 Clement, L., Kim-Sohn, K.A., Magnan, C., Kassis, N., Adnot, P., Kergoat, M., Assimacopoulos-Jeannet, F., Penicaud, L., Hsu, F., Turk, J. and Ktorza, A. (2002) Lipids 37, 501-506.

11 Han, X., Abendschein, D. R., Kelley, J.G. and Gross, R.W. (2000) Biochem. J. 352, 79–89.

12 Han, X. and Gross, R.W. (2003) J. Lipid Res. 44, 1071-1079.

13 Wang, C., Kong, H., Guan, Y., Yang, J., Gu, J., Yang, S. and Xu, G. (2005) Anal. Chem. 77, 4108-4116.

14 Xie, Y., Gibbs, T.C., Mukhin, Y.V. and Meier, K.E. (2002) J. Biol. Chem. 277, 32516-32526.

15 Sutphen, R., Xu, Y., Wilbanks, G.D., Fiorica, J., Grendys, E.C. Jr., LaPolla, J.P., Arango, H., Hoffman, M.S., Martino, M., Wakeley, K.,

Griffin, D., Blanco, R.W., Cantor, A.B., Xiao, Y.J. and Krischer, J.P. (2004) Cancer Epidemiol. Biomarkers Prev. 13, 1185-1191.
16 Han, X., Holtzman, D.M., McKeel, D.W. Jr., Kelley, J. and Morris, J.C. (2002) J. Neurochem. 82, 809-818.
17 Han, X., Fagan, A.M., Cheng, H., Morris, J.C., Xiong, C. and Holtzman, D.M. (2003) Ann. Neurol. 54, 115-119. Erratum in: (2003) Ann. Neurol. 54, 693.
18 DeLong, C.J., Baker, P.R., Samuel, M., Cui, Z. and Thomas, M.J. (2001) J. Lipid Res. 42, 1959-1968.
19 Fenn, J.B., Mann, M., Meng, C.K., Wong, S.F. and Whitehouse, C.M. (1989) Science 246, 64-71.
20 Fenn, J.B., Mann, M., Meng, C.K., Wong, S.F. and Whitehouse, C.M. (1990) Mass Spec. Rev. 9, 37-70.
21 Cech, N.B. and Enke, C.G. (2001) Mass Spec. Rev. 20, 362-387.
22 Cole, R.B. (1997) in Electrospray Ionization Mass Spectrometry Fundamentals, Instrumentation and Application John Wiley & Sons, Inc., New York, NY. pp. 1-174.
23 de la Mora, J.F., van Berkel, G., Enke, C.G., Cole, R.B., Martinez-Sanchez, M. and Fenn, J.B. (2000) J. Mass Spec. 35, 939-952.
24 Griffiths, W.J., Jonsson, A.P., Liu, S., Rai, D.K. and Wang, Y. (2001) Biochem. J. 355, 545-561.
25 Cole, R.B. (2000) J. Mass Spec. 35, 763-772.
26 Kebarle, P. (2000) J. Mass Spec. 35, 804-817.
27 Higashi, T., Takido, N., Yamauchi, A. and Shimada, K. (2002) Anal. Sci. 18, 1301-1307.
28 Hoffmann, E.D. (1996) J. Mass Spec. 31, 129-137.
29 Kuksis, A. and Myher, J.J. (1995) J. Chromatogr. B 671, 35-70.
30 Sullards, M.C., Lynch, D.V., Merrill, A.H., Jr. and Adams, J. (2000) J. Mass Spec. 35, 347-353.
31 Hsu, F.F. and Turk, J. (2000) J. Amer. Soc. Mass Spec. 11, 437-449.
32 Han, X. and Gross, R.W. (1995) J. Amer. Soc. Mass Spec. 6, 1202-1210.
33 Hsu, F.F., Bohrer, A. and Turk, J. (1998) J. Amer. Soc. Mass Spec. 9, 516-526.
34 Hsu, F.F., Bohrer, A. and Turk, J. (1998) Biochim. Biophys. Acta 1392, 202-216.
35 Taguchi, R., Houjou, T., Nakanishi, H., Yamazaki, T., Ishida, M., Imagawa, M. and Shimizu, T. (2005) J. Chromatogr. B 823, 26-36.
36 Ekroos, K., Chernushevich, I.V., Simons, K. and Shevchenko, A. (2002) Anal. Chem. 74, 941-949.
37 Han, X, Yang, J., Cheng, H., Ye, H. and Gross, R.W. (2004) Anal. Biochem. 330, 317-331.
38 Brügger, B., Erben, G., Sandhoff, R., Wieland, F.T. and Lehmann, W.D. (1997) Proc. Nat. Acad. Sci. U.S.A. 94, 2339-2344.
39 Han, X. and Gross, R.W. (2001) Anal. Biochem. 295, 88-100.
40 Han, X. and Gross, R.W. (2005) Mass Spec. Rev. 24, 367-412.
41 Welti, R., Wang, X. and Williams, T.D. (2003) Anal. Biochem. 314, 149-152.

42 Hsu, F.F., Turk, J., Rhoades, E.R., Russell, D.G., Shi, Y. and Groisman, E.A. (2005) J. Amer. Soc. Mass Spec. 16, 491-504.

43 Wenk, M.R., Lucast, L., DiPaolo, G., Romanelli, A.J., Suchy, S.F., Nussbaum, R.L., Cline, G.W., Shulman, G.I., McMurray, W. and DeCamilli, P. (2003) Nat. Biotech. 21, 813-817.

44 Milne, S.B., Ivanova, P.T., DeCamp, D., Hseuh, R.C., and Brown, H.A. (2005) J. Lipid Res. 46, 1796-1802.

45 Merrill, A.H. Jr., Sullards, M.C., Allegood, J.C., Kelly, S. and Wang, E. (2005) Methods 36, 207-224.

46 Han, X. (2002) Anal. Biochem. 302, 199-212.

47 Han, X. and Cheng, H. (2005) J. Lipid Res. 46, 163-175.

48 Tsui, Z.-C., Chen, Q.-R., Thomas, M.J., Samuel, M. and Cui, Z. (2005) Anal. Biochem. 251-258.

49 Gage, D.A., Huang, Z.H. and Benning, C. (1992) Lipids 27, 632-636.

50 Kim, Y.H., Choi, J.-S., Yoo, J.S., Park, Y.-M., and Kim, M.S. (1999). Anal. Biochem. 267, 260-270.

51 Isaac, G., Bylund, D., Månsson, J.-E., Markides, K.E. and Bergquist, J. (2003) J. Neurosci. Methods 128, 111-119.

52 Fridriksson, E.K., Shipkova, P.A., Sheets, E.D., Holowka, D., Baird, B. and McLafferty, F.W. (1999). Biochemistry 38, 8056-8063.

53 Ivanova, P.T., Cerda, B.A., Horn, D.M., Cohen, J.S., McLafferty, F.W. and Brown, H.A. (2001) Proc. Nat. Acad. Sci. U. S. A. 98, 7152-7157.

54 Wanjie, S.W., Welti, R., Moreau, R.A. and Chapman, K.D. (2005). Lipids 40, 773-785.

55 Bayer, E., Gfrörer, P. and Rentel, C. (1999) Angew. Chem. Int. Ed. 38, 992-995.

56 Kingsley, P.J. and Marnett, L.J. (2003) Anal. Biochem. 314, 8-15.

57 Thompson, R.Q., Phinney, K.W., Welch, M.J. and White V.E. (2005) Anal. Bioanal. Chem. 381, 1441-1451.

58 Medvedovici, A., Lazou, K., d'Oosterlinck, A., Zhao, Y. and Sandra, P. (2002) J. Sep. Sci. 25, 173-178.

59 Strohschein, S., Rentel, C., Lacker, T., Bayer, E. and Albert, K. (1999) Anal. Chem. 71, 1780-1785.

60 Al-Talla, Z.A. and Tolley, L.T. (2005) Rapid Commun. Mass Spec. 19, 2337-2342.

61 Gaub, M., von Brocke, A., Roos, G. and Kovar, K.A. (2004) Phytochem. Anal. 15, 300-305.

62 Sandra, P., Medvedovici, A., Zhao, Y. and David, F. (2002) J. Chromatogr. A 974, 231-241.

63 Havrilla, C.M., Hachey, D.L. and Porter, N. (2000) J. Amer. Chem. Soc. 122, 8042–8055.

64 Milne, G.L. and Porter, N.A. (2001) Lipids 36, 1265-1275.

65 Seal, J.R., Havrilla, C.M., Porter, N.A. and Hachey, D.L. (2003) J. Amer. Soc. Mass Spec. 14, 872-880.

66 Seal, J.R. and Porter, N.A. (2004) Anal. Bioanal. Chem. 378, 1007-1013. Erratum in: Anal. Bioanal. Chem. 380, 356.

67 Yin, H. and Porter, N.A. (2005) Antiox. Redox Signal. 7, 170-184.

68 Cui, M., Song, F., Liu, Z. and Liu, S. (2001), Rapid Commun. Mass Spec. 15, 586-595.
69 Ho, Y.-P., Huang, P.-C. and Deng, K.-H. (2003) Rapid Commun. Mass Spec. 17, 114-121.
70 Zavitsanos, P. (2003) in: Liquid Chromatography/Mass Spectrometry, MS/MS and Time of Flight MS. Analysis of Emerging Contaminants (I. Ferrer and E.M. Thurman, eds.) American Chemical Society, Washington, DC, pp. 207-237.
71 Diaz-Cruz, M.S., de Alda, M.J.L., Lopez, R. and Barcelo, D. (2003) J. Mass Spec. 38, 917-923.
72 Mezine, I., Zhang, H., Macku, C. and Lijana, R. (2003) J. Agric. Food Chem. 51, 5639-5646.
73 Sanchez-Machado, D.I., Lopez-Hernandez, J., Paseiro-Losada, P. and Lopez-Cervantes, J. (2004) Biomed. Chromatogr. 18, 183-190.
74 Higashi, T. and Shimada, K. (2004) Bunseki Kagaku 53, 645-655.
75 Syage, J.A., Hanold, K.A., Lynn, T.C., Horner, J.A. and Thakur, R.A. (2004) J. Chromatogr. A 1050, 137-149.
76 Higashi, T., Takayama, N. and Shimada, K. (2005) J. Pharmaceut. Biomed. 39, 718-723.
77 Palmgren, J.J., Toyras, A., Mauriala, T., Monkkonen, J. and Auriola, S. (2005) J. Chromatogr. B 821, 144-152.
78 Raith, K., Brenner, C., Farwanah, H., Muller, G., Eder, K. and Neubert, R.H.H. (2005) J. Chromatogr. A 1067, 207-211.
79 Nagy, K., Bongiorno, D., Avellone, G., Agozzino, P., Ceraulo, L. and Vekey, K. (2005) J. Chromatogr. A 1078, 90-97.
80 Byrdwell, W.C. (2001) Lipids 36, 327-346.
81 Byrdwell, W.C., Emken, E.A., Neff, W.E. and Adlof, R.O. (1996) Lipids 31, 919-935.
82 Byrdwell, W.E., Neff, W.E. and List, G.R. (2001) J. Agric. Food Chem. 49, 446-457.
83 Byrdwell, W.C. and Neff, W.E. (2002) Rapid Commun. Mass Spec. 16, 300-319.
84 Byrdwell, W.C. and Neff, W.E. (2004) J. Amer. Oil Chem. Soc. 81, 13-26.
85 van Vliet, M.H. and van Kempen, G.M.P. (2004) Eur. J. Lipid Sci. Technol. 106, 697-706.
86 Momchilova, S., Tsuji, K., Itabashi, Y. and Nikolova-Damyanova, B. (2004) J. Sep. Sci. 27, 1033-1036.
87 Fauconnot, L., Hau, J., Aeschlimann, J.-M., Fay, L.-B. and Dionisi, F. (2004) Rapid Commun. Mass Spec. 18, 218-224.
88 Cheng, C., Gross, M.L. and Pittenauer, E. (1998) Anal. Chem. 70, 4417-4426.
89 Hsu, F.-F. and Turk, J. (1999) J. Amer. Soc. Mass Spec. 10, 587-599.
90 Schiller, J., Sub, R., Arnhold, J., Fuchs, B., Lebig, J., Muller, M., Petkovic, M., Spalteholz, H., Zschornig, O. and Arnold, K. (2004) Prog. Lipid Res. 43, 449-488.
91 Harvey, D.J. (2003) Int. J. Mass Spec. 226, 1-35.
92 Harvey, D.J. (1995) J. Mass Spec. 30, 1311-1324.

93 van Baar, B.L. (2000) FEMS Microbiol. Rev. 24, 193-219.
94 Calvano, C.D., Palmisano, F. and Zambonin, C.G. (2005) Rapid Commun. Mass Spec. 19, 1315-20.
95 Jones, J.J., Batoy, S.M.A.B. and Wilkins, C.L. (2005) Comput. Biol. Chem. 29, 294-302.
96 Jones, J.J., Stump, M.J., Fleming, R.C., Lay, J., Jackson O. and Wilkins, C.L. (2004) J. Amer. Soc. Mass Spec. 15, 1665-1674.
97 Jones, J.J., Stump, M.J., Fleming, R.C., Lay, J., Jackson O. and Wilkins, C.L. (2003) Anal. Chem. 75, 1340-1347.
98 Ivleva, V.B., Elkin, Y.N., Budnik, B.A., Moyer, S.C., O'Connor, P.B. and Costello, C.E. (2004) Anal. Chem. 76, 6484-6491.
99 URL: http://www.lipidmaps.org/about/researchplan.html.
100 Jackson, S.N., Wang, H.-Y.J. and Woods, A.S. (2005) Anal. Chem. 77, 4523-4527.
101 Koivusalo, M., Haimi, P., Heikinheimo, L., Kostiainen, R. and Sommerharju, P. (2001) J. Lipid Res. 42, 663-672.
102 Blom, T.S., Koivusalo, M., Kuismanen, E., Kostiainen, R., Sommerharju, P. and Ikonen, E. (2001) Biochemistry 40, 14635-14644.
103 Muller, M., Schiller, J., Petkovic, M., Oehrl, W., Heinze, R., Wetzker, R., Arnold, K. and Arnhold, J. (2001) Chem. Phys. Lipids 110, 151-164.
104 Zollner, P., Schmid, E.R. and Allmaier, G. (1996) Rapid Commun. Mass Spec. 10, 1278-1282.
105 Gobey, J., Cole, M., Janiszewski, J., Covey, T., Chau, T., Kovarik, P. and Corr, J. (2005) Anal. Chem. 77, 5643-5654.
106 Woods, A.S., Ugarov, M., Egan, T., Koomen, J., Gillig, K.J., Fuhrer, K., Gonin, M. and Schultz, J.A. (2004) Anal. Chem. 76, 2187-2195.
107 Hunnam, V., Harvey, D.J., Priestman, d.A., Bateman, R.H., Bordoli, R.S. and Tyldesley, R. (2001) J. Amer. Soc. Mass Spec. 12, 1220-1225.
108 Chaurand, P., Stoeckli, M. and Caprioli, R.M. (1999) Anal. Chem. 71, 5263-5270.
109 Chaurand, P., Schwartz, S.A. and Caprioli, R.M. (2002) Curr. Opin. Chem. Biol. 6, 676-681.
110 Chaurand, P., Sanders, M.E., Jensen, R.A. and Caprioli, R.M. (2004) Amer. J. Pathol. 165, 1057-1068.
111 Chaurand, P., Schwartz, S.A., Billheimer, D., Xu, B.J., Crecelius, A. and Caprioli, R.M. (2004) Anal. Chem. 76, 1145-1155.
112 Stoeckli, M., Chaurand, P., Hallahan, D.E. and Caprioli, R.M. (2001) Nat. Med. 7, 493-496.
113 Todd, P.J., Schaaff, T.G., Chaurand, P. and Caprioli, R.M. (2001) J. Mass Spec. 36, 355-369.
114 Griesser, H.J., Kingshott, P., McArthur, S.L., McLean, K.M., Kinsel, G.R. and Timmons, R.B. (2004) Biomaterials 25, 4861-4875.
115 Caldwell, R.L. and Caprioli, R.M. (2005) Mol. Cell. Proteomics 4, 394-401.
116 Reyzer, M.L. and Caprioli, R.M. (2005) J. Proteome Res. 4, 1138-1142.
117 Rubakhin, S.S., Jurchen, J.C., Monroe, E.B. and Sweedler, J.V. (2005) Drug Discov. Today 10, 823-837.

118 Pacholski, M.L. and Winograd, N. (1999) Chem. Rev. 99, 2977-3005.
119 Belu, A.M., Graham, D.J. and Castner, D.G. (2003) Biomaterials 24, 3635-3653.
120 Paul, A.J. (2005) Spectrosc. Eur. 17, 25-27.
121 Brunelle, A., Touboul, D. and Laprevote, O. (2005) J. Mass Spec. 40, 985-999.
122 Guerquin-Kern, J.-L., Wu, T.-D., Quintana, C. and Croisy, A. (2005) Biochim. Biophys. Acta 1724, 228-238.
123 Sodhi, R.N. (2005) Analyst 129, 483-487.
124 Rujoi, M., Estrada, R. and Yappert, M.C. (2004) Anal. Chem. 76, 1657-1663.
125 Touboul, D., Piednoël, H., Voisin, V., De La Porte, S., Brunelle, A., Halgand, F. and Laprévote, O. (2004) Eur. J. Mass Spec. 10, 657-664.
126 Sluszny, C., Yeung, E.S. and Nikolau, B.J. (2005) J. Amer. Soc. Mass Spec. 16, 107-115.
127 Jackson, S.N., Wang, H.Y. and Woods, A.S. (2005) J. Amer. Soc. Mass Spec. 16, 2052-2056.
128 Jackson, S.N., Wang, H.Y., Woods, A.S., Ugarov, M., Egan, T. and Schultz, J.A. (2004) J. Amer. Soc. Mass Spec. 16, 133-138.
129 Tempez, A., Ugarov, M., Egan, T., Schultz, J.A., Novikov, A., Della-Negra, S., Lebeyec, Y., Pautrat, M., Caroff, M., Smentkowski, V.S., Wang, H.Y., Jackson, S.N. and Woods, A.S. (2005) J. Proteome Res. 4, 540-545.
130 Touboul, D., Brunelle, A., Halgand, F., De La Porte, S. and Laprévote, O. (2005) J. Lipid Res. 46, 1388-1395.
131 Nygren, H., Malmberg, P., Kriegeskotte, C. and Arlinghaus, H.F. (2004) FEBS Lett. 566, 291-293.
132 Nygren, H., Börner, K., Hagenhoff, B., Malmberg, P. and Månsson, J.-E. (2005) Biochim. Biophys. Acta 1737, 102-110.
133 Sjövall, P., Lausmaa, J., Nygren, H., Carlsson, L. and Malmberg, P. (2003) Anal. Chem. 75, 3429-3434.
134 Sjövall, P., Lausmaa, J. and Johansson, B. (2004) Anal. Chem. 76, 4271-4278.
135 Roddy, T.P., Cannon, D.M. Jr., Meserole, C.A., Winograd, N. and Ewing, A.G. (2002) Anal. Chem. 74, 4011-4019.
136 Roddy, T.P., Cannon, D.M. Jr., Ostrowski, S.G., Winograd, N. and Ewing, A.G. (2002) Anal. Chem. 74, 4020-4026.
137 Ostrowski, S.G., Van Bell, C.T., Winograd, N. and Ewing, A.G. (2004) Science 305, 71-73.
138 Ingram, J.C., Bauer, W.F., Lehman, R.M., O'Connell, S.P. and Shaw, A.D. (2003) J. Microbiol. Methods 53, 295-307.
139 Amemiya, T., Tozu, M. and Ohashi, Y. (2004) J. J. Ophthalmol. 48, 287-293.
140 Perkins, M.C., Roberts, C.J., Briggs, D., Davies, M.C., Friedmann, A., Hart, C.A. and Bell, G.A. (2005) Planta 221, 123-134.
141 Imai, T., Tanabe, K., Kato, T. and Fukushima, K. (2005) Planta 221, 549-556.

142 Ostrowski, S.G., Szakal, C., Kozole, J., Roddy, T.P., Xu, J., Ewing, A.G. and Winograd, N. (2005) Anal. Chem. 77, 6190-6196.
143 Touboul, D., Halgand, F., Brunelle, A., Kersting, R., Tallarek, E., Hagenhoff, B. and Laprévote, O. (2004) Anal. Chem. 76, 1550-1559.
144 Nygren, H. and Malmberg, P. (2004) J. Microsc. 215, 156-161.
145 Van Berkel, G.J., Sanchez, A.D. and Quirke, J.M. (2002) Anal. Chem. 74, 6216-6223.
146 Ford, M.J. and Van Berkel, G.J. (2004) Rapid Commun. Mass Spec. 18, 1303-1309.
147 Takáts, Z., Wiseman, J.M., Gologan, B. and Cooks, R.G. (2004) Science 306, 471-473.
148 Takáts, Z., Cotte-Rodriguez, I., Talaty, N., Chen, H. and Cooks, R.G. (2005) Chem. Commun. 15, 1950-1952.
149 Takáts, Z., Wiseman, J.M. and Cooks, R.G. (2005) J. Mass Spec. 40, 1261-1275.
150 Talaty, N., Takáts, Z. and Cooks, R.G. (2005) Analyst 130, 1624-1633.
151 Ford, M.J., Deibel, M.A., Tomkins, B.A. and Van Berkel, G.J. (2005) Anal. Chem. 77, 4385-4389.
152 Ford, M.J., Kertesz, V. and Van Berkel, G.J. (2005) J. Mass Spec. 40, 866-875.
153 Van Berkel, G.J., Ford, M.J. and Deibel, M.A. (2005) Anal. Chem. 77, 1207-1215.
154 Asano, K.G., Ford, M.J., Tomkins, B.A. and Van Berkel, G.J. (2005) Rapid Commun. Mass Spec. 19, 2305-2312.
155 McEwen, C.N., McKay, R.G. and Larsen, B.S. (2005) Anal. Chem. 77, 7826-7831.
156 Cody, R.B., Laramée, J.A. and Durst, H.D. (2005) Anal. Chem. 77, 2297-2302.
157 Wiseman, J.M., Puolitaival, S.M., Takáts, Z., Cooks, R.G. and Caprioli, R.M. (2005) Angew. Chem. Int. Ed. 44, 7094-7097.
158 Gao, S.M., Zhang, Z.P. and Karnes, H.T. (2005) J. Chromatogr. B Analyt. Technol. Biomed. Life Sci. 825, 98-110.
159 Halket, J.M. and Zaikin, V.G. (2003) Eur. J. Mass Spec. 9, 1-21.
160 Halket, J.M. and Zaikin, V.G. (2004) Eur. J. Mass Spec. 10, 1-19.
161 Halket, J.M. and Zaikin, V.G. (2005) Eur. J. Mass Spec. 11, 127-160.
162 Halket, J.M., Waterman, D., Przyborowska, A.M., Patel, R.K., Fraser, P.D. and Bramley, P.M. (2005) J. Exp. Bot. 56, 219-243.
163 Wells, R.J. (1999) J. Chromatogr. A 843, 1-18.
164 Zaikin, V.G. and Halket, J.M. (2003) Eur. J. Mass Spec. 9, 421-434.
165 Zaikin, V.G. and Halket, J.M. (2004) Eur. J. Mass Spec. 10:555-568.
166 Zaikin, V.G. and Halket, J.M. (2005) Eur. J. Mass Spec. 11:611-636.
167 Carrapiso, A.I. and Garcia, C. (2000) Lipids 35, 1167-1177.
168 Rosenfeld, J.M. (2002) Anal. Chim. Acta 465, 93-100.
169 Wang, Y., Krull, I.S., Liu, C. and Orr J.D. (2003) J. Chromatogr. B Biomed. Sci. Appl. 793, 3-14.
170 Higashi, T. and Shimada, K. (2004) Anal. Bioanal. Chem. 378, 875-882.

171 Han, X., Yang, K., Cheng, H., Fikes, K.N. and Gross, R.W. (2005) J. Lipid Res. 46, 1548-1560.
172 Berdyshev, E.V., Gorshkova, I.A., Garcia, J.G., Natarajan, V. and Hubbard, W.C. (2005) Anal. Biochem. 339, 129-136.
173 Lee, S.H., Williams, M.V., DuBois, R.N. and Blair, I.A. (2003) Rapid Commun. Mass Spec. 17, 2168-2176.
174 Han, X. and Gros, R.W. (2001) Anal. Biochem. 295, 88-100.
175 Han, X. and Gross, R.W. (2003) J. Lipid Res. 44, 1071-1079.
176 Liebisch, G., Lieser, B., Rathenberg, J., Drobnik, W. and Schmitz, G.. (2004) Biochim. Biophys. Acta. 1686, 108-117. Erratum in: (2005) Biochim. Biophys. Acta. 1734, 86-89.
177 Kopka, J., Schauer, N., Krueger, S. Birkmeyer, C., Usadel, B., Bergmüller, E., Dörmann, P., Weckwerth, W., Gibon, Y., Stitt, M., Willmitzer, L., Fernie, A.R. and Steinhauser, D. (2005) Bioinformatics 21, 1635-1638.
178 Ivanova, P.T., Milne, S.B., Forrester, J.S., and Brown, H.A. (2004) Mol. Interv. 4, 86-96.
179 Forrester, J.S., Milne, S.B., Ivanova, P.T. and Brown, H.A..(2004) Mol. Pharmacol. 65, 813-821.
180 Fang, J., Barcelona, M.J. and Semrau, J.D. (2000) FEMS Microbiol. Letts. 189, 67-72.
181 Mortuza, G.B., Neville, W.A., Delaney, J., Waterfield, C.J. and Camilleri, P. (2003) Biochim. Biophys. Acta. 1631, 136-146.
182 Davidov, E., Clish, C.B., Oresic, M., Meys, M., Stochaj, W., Snell, P., Lavine, G., Londo, T.R., Adourian, A., Zhang, X., Johnston, M., Morel, N., Marple, E.W., Plasterer, T.N., Neumann, E., Verheij, E., Vogels, J.T., Havekes, L.M., van der Greef, J. and Naylor, S. (2004) OMICS 8, 267-288.

PAIRED-END GENOMIC SIGNATURE TAGS: A METHOD FOR THE FUNCTIONAL ANALYSIS OF GENOMES AND EPIGENOMES

John J. Dunn, Sean R. McCorkle, Logan Everett and Carl W. Anderson

Biology Department
Brookhaven National Laboratory
Upton, NY 11973-5000

INTRODUCTION

Interactions between eukaryotic transcription factors and their cognate DNA binding sites form fundamental networks within cells that control critical steps during development and tissues-specific gene expression. These interactions also are important in regulating cellular responses to stresses, and their dysfunction contributes to numerous diseases. Therefore, determining the *in vivo* genome-wide binding distribution of transcription factors is an important step towards developing an understanding of the regulatory networks in a living cell as well as their changes in response to specific stimuli. Methods based on chromatin immunoprecipitation (ChIP) are beginning to provide an increasing detailed view of these dynamic events. This assay was originally developed to monitor histone modifications and then modified to detect binding of specific transcription factors to native chromatin (1-3). In this method, transcription factors are reversibly cross-linked to their binding sites using formaldehyde to freeze intracellular protein-DNA complexes, the DNA is sonicated to generate fragments with lengths

Genetic Engineering, Volume 28, Edited by J. K. Setlow

of ~500 to ~2,000 bp, and individual transcription factor-DNA complexes are immunoprecipitated using specific antibodies to the native protein or to a suitable epitope fused in-frame to the target protein's coding sequence. The DNA fragments enriched by ChIP can be identified by a variety of means such as cloning or amplification with gene specific primers, hybridization to microarrays containing subsets of the genomic sequences or by Serial Analysis of Gene Expression (SAGE)-type approaches (4) that extract short sequence identifier tags or Genomic Signature Tags (GSTs) from the ChIP DNA and then use this information to map the DNA back to the genome. In this article we will review the basic steps for generating GSTs, their application to analysis of ChIP data and will introduce several modifications of our original SACO *(for Serial Analysis of Chromatin* Occupancy) method (5) that can simultaneously generate tags from both ends of the ChIP fragments and preserve their spatial relationship to each other. The sequences of each tag in combination uniquely identify the region of the genome from which the original SACO fragment was derived and encompass the sequence of the site to which the transcription factor was bound.

This same approach can also be used to obtain paired sequence tags from the ends of any DNA fragment. At the whole genome level any changes in the resulting paired-end profile can provide a sensitive method for distinguishing between closely-related genomes or genomes that have undergone deletions, insertions or other rearrangements that cause the appearance of new diTAG pairs. Detection and characterization of discrepancies between observed diTAG pairs from reference and test genomes can, in principle, detect structural variations with the same precision as afforded by paired-end sequencing of fosmid or bacterial-artificial chromosome libraries (6). Such changes are characteristic of many cancers as are changes in CpG methylation in CpG islands, which are clusters of CpG dinucleotides that are found in front of about half of human genes (7). Methylation of cytosine within these islands caused inhibition of downstream gene expression, and aberrant methylation is an important mechanism for gene activation or inactivation in cancer. In this article, we briefly review how paired-end diTAGs can be obtained from DNA fragments associated with methylated CpG islands.

WHAT ARE GENOMIC SIGNATURE TAGS?

Genomic Signature Tags (GSTs) are the products of a method we developed for identifying and qualitatively analyzing genomic DNAs (8). Two major principles underlie this method: first, short DNA sequences (18-21 bp) are sufficient to identify unique sites within a genome; second, concatenation of these short DNA sequences, as in SAGE (9), greatly increases sequence throughput. The original GST method begins with cutting the DNA sample with a type II restriction enzyme, also termed the fragmenting enzyme, to produce fragments with cohesive ends. After digestion with the first enzyme, the cohesive ends are biotinylated, and the sample is digested with *Nla* III, also called the anchoring enzyme, which cleaves leaving 4 base cohesive ends. Since *Nla* III has a 4 bp recognition sequence (CATG), it theoretically cleaves on average every 256 bp, and nearly every fragment in the original digest will be cleaved at least once to produce two biotinylated end fragments which are recovered by binding to streptavidin-coated magnetic beads. The bound DNA fragments are then ligated with a linker cassette that creates partially overlapping *Mme* I (TCCRAC) and *Nla* III (CATG) recognition sites; i.e., TCCRACATG with the C in bold being shared by both recognition sequences. *Mme* I is a type IIS restriction enzyme, with cut sites 20-21/18-19 bp past its recognition site. Cutting the linkered DNA with *Mme* I releases the linker and 17-18/15-16 bp immediately 3′ to the *Nla* III site. These CATG+17 or 18 bp sequences become the identifier tags which are PCR amplified and ligated together to form ≥500 bp long concatemers prior to cloning and DNA sequencing. Because each clone contains multiple tags, sequencing throughput increases accordingly.

SERIAL ANALYSIS OF CHROMATIN OCCUPANCY (SACO)

In principle, *Mme* I derived tags can be used to identify the region of the genome from which any DNA or RNA (after conversion to cDNA) fragment is derived. As shown in Figure 1, *in silico* simulations of tag uniqueness *vs.* tag

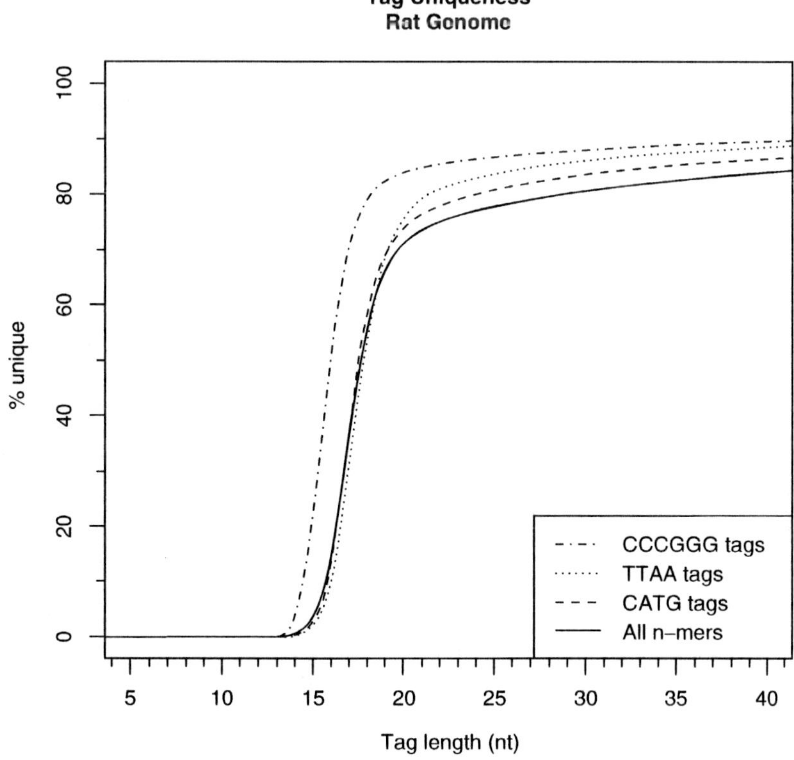

Figure 1. Plot of tag uniqueness *vs.* tag length for the entire rat genome. Plotted is tag length, which includes nucleotides at fragment ends specified by restriction sites: *Mse* I-TTAA; *Nla* III-CATG; *Sma* I CCCGGG or random ends (all N-mers) derived by sonication.

length for the rat genome show that uniqueness rapidly increases for lengths longer than ~12 bp and is limited only by the presence of highly repetitive regions in genomes. Similar profiles are obtained for the mouse and human genomes. With these as background, we reasoned that a ChIP-to-tag sequencing approach could be used to identify the genomic locations of ChIP-derived DNA fragments. To establish the effectiveness of the method, we set out to map globally the c-AMP response element binding protein (CREB) binding sites in the genome of rat PC12 cells (5). CREB was known to bind the cAMP-response element (CRE) (TGACGTCA) present in the promoters of many inducible genes (reviewed in 10). To increase the chances that CREB would be associated with CRE sequences, we first incubated the cells with forskolin to activate the enzyme adenylcyclase and increase the intracellular levels of cyclic AMP. The cells were then treated with formaldehyde, and, after randomly fragmenting the entire genome by sonication, the samples were subjected to ChIP using an anti-CREB antibody or, as a control, non-specific IgG. Real-time quantitative PCR showed that the CREB antibody provided an ~100-fold enrichment for c-fos (and other CREB targets) in the immunoprecipitates as compared to the IgG control. The ends of the CREB ChIP DNA were polished (protruding 3′ and 5′ ends were made flush by incubation with *E. coli* (Klenow fragment) and T4 DNA polymerases plus all four deoxynucleotide triphosphates and ligated to adapters for limited PCR amplification using biotinylated adapter-specific primers). The resulting DNA was digested with *Nla* III, and a modified Long-SAGE procedure was used to create concatemerized chains of randomly-associated 21 bp GSTs which were then cloned and sequenced. We termed this approach SACO; to demonstrate its utility, the sequences of ~75,000 tags from the PC12-derived library were determined. More than 40,000 CREB-SACO tags that mapped to unique loci in the rat genome were identified; 6,302 of these were identified two or more times. When these data were integrated with sequence annotation maps of the rat genome, forty percent of these loci were within 2 kb of the transcriptional start site of an annotated gene, and 72% were within 1 kb of a putative cAMP response element. In addition, CREB binding was confirmed for all loci supported by multiple tag hits (53 of 53 that were tested), and many of these loci were located upstream from genes not previously known to be regulated by CREB. These included genes for transcriptional regulators, chromatin modifying enzymes, coactivators, and co-repressors. A surprising result of the CREB SACO study was that CREB binding sites were commonly located in bi-directional promoters. Thus, the CRE that controls c-fos expression, for example, also regulates expression of a noncoding RNA transcribed in the opposite direction (5).

Since publication of the SACO method, several papers have appeared that utilized similar approaches, attesting to the overall utility of tag-to-genome mapping of ChIP DNA fragments (11-14). In all of these procedures the tags, whether they are generated from an internal restriction site or directly from the 5′ and 3′ ends of the sonicated ChIP DNA fragments, are analyzed separately as independent bits of sequence data. When mapped correctly to the genome sequence, these tags locate within about 1 to 2 kb the site that was cross-linked *in vivo* to the immunoprecipitated protein. In practice finding these sites involves scanning the genome sequence in both directions from a tag's location for a nearby binding motif. The distance scanned is usually set at around twice the upper limit of the size of the ChIP DNA since when tags are analyzed separately, it is not known where they originated in the fragment, i.e., were they close to an end or more towards the middle of the ChIP fragment. To overcome this limitation, a new cloning strategy was developed by Ng and co-workers (15) that covalently links the tag sequences from each end of a DNA fragment into a paired diTAG structure. This approach, which was originally developed for identifying simultaneously both ends of full-length cDNAs, can also be used to map ChIP fragments with high precision.

PAIRED-END GENOMIC SIGNATURE TAGS (PE-GST)

The first step in the procedure is cloning of the DNA fragments into a special vector, pBEST (Both End Signature Tags), which is based on the pSCANS vector developed at BNL (http://genome.bnl.gov/Vectors/pscans.php). This low-copy number vector, with an isopropyl-beta-D-thiogalactopyranoside (IPTG) inducible origin of replication, was modified for efficient cloning of single DNA fragments in a manner that places them immediately adjacent to oppositely oriented *Mme* I recognition sequences (Figure 2). These are the only *Mme* I sites in the vector. Two *Bbs* I sites were placed between the *Mme* I sites in opposite orientations such that when the vector is cut with *Bbs* I, the linearized vector DNA will have non-self-ligatable ends with 4 nt overhangs (5′-GTCG-3′). A synthetic

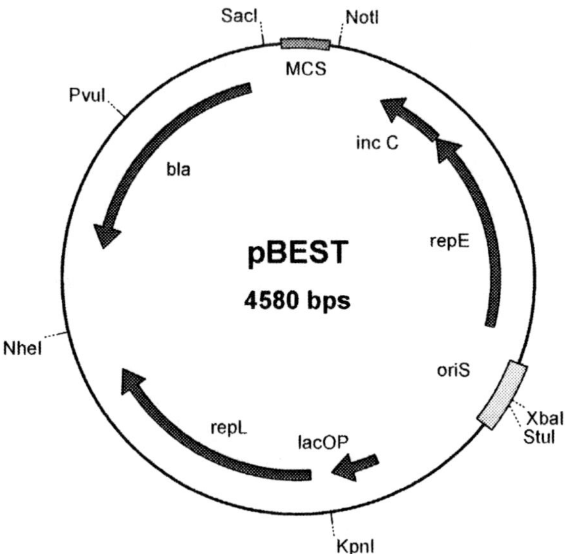

Figure 2a. Schematic diagram of pBEST paired-end vector. *ori*S, *rep*E and *inc* C are from the *E. coli* F factor, *lac*OP is the wild-type lac promoter, *rep*L is the lytic origin of replication from bacteriophage P1, and *bla* encodes β-lactamase activity (*amp*R). Several of the plasmid's unique restriction sites are indicated. MCS represents the cloning region, which is shown in greater detail in Figure 2b.

```
GACATACGAT TTAGGTGACA CTATAGAACT CTAATACGAC TCACTATAGG GAATTTGGCC
CTGTATGCTA AATCCACTGT GATATCTTGA GATTATGCTG AGTGATATCC CTTAAACCGG
   >>........SP6P.........>>
                                      >>.......T7P........>>

                                 MmeI                                BamHI
       BseRI        BamHI           BbsI       EcoRI       BbsI       MmeI
CTCGAGAGGA GCCAGGATCC GACTTGTCTT CACGAATTCA CGAAGACCAG TCGGAGGATC
GAGCTCTCCT CGGTCCTAGG CTGAACAGAA GTGCTTAAGT GCTTCTGGTC AGCCTCCTAG

 BseRI
CCTCCTCGCG GCCGCACGCG TACCCATAAT ACCCATAATA GCTGTTTGCC ATCGCGTATG
GGAGGAGCGC CGGCGTGCCG ATGGGTATTA TGGGTATTAT CGACAAACGG TAGCGCATAC

CATCGATCAC GTGTCCACGT TCTTTAATAG TGGACTCTTG TTCCAAACTG GAACAACACT
GTAGCTAGTG CACAGGTGCA AGAAATTATC ACCTGAGAAC AAGGTTTGAC CTTGTTGTGA
                                     REVERSE-PRIMER <<.............<

CGGATCGATC CGGCGCGCAC CGTGGGAAAA ACTCCAGGTA GAGGTACACA CGCGGATAGC
GCCTAGCTAG GCCGCGCGTG GCACCCTTTT TGAGGTCCAT CTCCATGTGT GCGCCTATCG
<<< Reverse-primer
```

Figure 2b. MCS region of pBEST used for producing diTAGs. The locations of the relevant restriction enzyme recognition sites are indicated as are those for the primers used to PCR amplify the diTAG concatemers. Complete cutting with *Bbs* I generates a linearized vector with 5′-GTCG-3′ overhangs (indicated).

double-stranded DNA cassette is used to append simultaneously *BtgZ* I and *Bbs* I recognition sites to the ends of blunt-ended ChIP DNAs. The bottom strand of the cassette is 5′ phosphorylated (p) and its 3′ end is amino modified to prevent self-ligation higher than dimers.

Cassette #1

```
                                            BtgZ I       Bbs I
5′          TCCGGTCTAC  TGAATTCCGA  ACGCGATGCT  GAAGACCACG  AC
3′    Amino-AGGCCAGATG  ACTTAAGGCT  TGCGCTACGA  CTTCTGGTGC  TGp
```

Similar cassettes with appropriate overhangs are used if dealing with fragments with cohesive ends. Cutting these cassettes with either *BtgZ* I or *Bbs* I generates 4 bp overhangs (5′-CGAC-3′) on the ends of the linkered DNA that are complementary to the overhangs of the *Bbs* I cut vector. After overnight ligation with excess linker, the ligation products are purified on a Qiagen Qiaquick PCR purification column, and the eluant is PCR amplified using 5′-biotin-TCCGGTCTACTGAATTCCGAAC-3′ as primer. Ideally one should set up several different PCR reactions varying the amount of input template and PCR cycles. Amplified material should then be analyzed by agarose gel electrophoresis. The products should produce a smear that is similar in its size range to the DNA fragments in the sonicated ChIP starting material. The appropriate samples are phenol-chloroform extracted; then a portion is digested with *BtgZ* I and a similar portion with *Bbs* I to minimize loss of fragments with internal *BtgZ* I or *Bbs* I sites. After digestion the samples are combined, the cleaved linker cassettes are removed by gel electrophoresis or by binding to streptavidin beads, and the ChIP fragments are ligated into *Bbs* I cut pBEST to generate recombinant plasmids with only single inserts. These are then electroporated into *E. coli* D1210, and the library is plated on ZYM5052 plus ampicillin (50 μg/ml) plates (16). Growth on ZYM5052 agar provides for solid-phase plasmid amplification without having to add IPTG to the plates, which should maintain library representation better than growth in liquid culture. The number of colonies required at this stage is determined by the estimated number of targets in the genome being investigated; we routinely target 1-10 X 10^5 cfu as a convenient benchmark. Cells can be plated at a density just below what is needed to provide for a confluent lawn. After overnight 37°C incubation, the resultant lawn of bacterial colonies is harvested by scraping into several ml of liquid medium and pelleted by centrifugation. Plasmid DNA preparation is performed e.g., by using a Qiagen Tip500 kit.

These clones now contain an *Mme* I site (TCCGAC) on each side of the DNA insert oriented so that digestion with *Mme* I cleaves 20-21 bp into the inserts from both their 5′ and 3′ ends. Consequently, despite the variable sizes of the original inserts, the vector-plus the two 20-21 bp tags on each end will be of a constant size (approx. 4,500 bp) that can be easily recognized upon agarose gel electrophoresis and can be purified from the unwanted internal ChIP-derived fragments that are produced during the *Mme* I digestion step. Approximately 5-10 μg of plasmid DNA is digested using *Mme* I as per the manufacturer's conditions (NEB), and the entire digestion reaction is then electrophoresed on a 0.7% low melt agarose gel. After staining, the vector plus tags band is excised, and the DNA is recovered. These molecules will eventually be ligated under dilute conditions to form circles that bring the tags at each end physically adjacent to each other as paired-end diTAGs. However, at this stage only 1 in 16 of the overhangs left following *Mme* I digestion are expected to be complementary to each other and able to form monomeric circles. Therefore, they either have to be removed by the 3′ exonuclease activity of T4 DNA polymerase prior to blunt end ligation or, alternatively, ligated with a special DNA adapter cassette with a 16-fold degenerate two-base 3′ overhang, which makes it compatible with all possible 3′ overhangs generated by *Mme* I digestion. Plasmid maps and detailed protocols are available on our web site (http://genome.bnl.gov/pBEST).

We initially used a blunt-ending approach to analyze DNA sequences associated with the product of the human p53 tumor suppressor gene (*TP53*). This 393 amino acid long polypeptide is known to function as a homotetrameric, sequence-specific transcription factor controlling cell cycle progression, DNA repair, and the induction of apoptosis and senescence in response to a variety of genotoxic and non-genotoxic stress signals (17-20). Genomic studies have shown that p53 induces or inhibits the expression of more than 1,500 human genes, but only a handful of p53 response elements (p53REs) have been characterized. The p53 tetramer binds a consensus DNA sequence, 5′-RRRCWWGYYY(N = 0-14) RRRCWWGYYY -3′, which consists of pairs of inverted repeats separated by 0 to 14 bp to create a 20 bp binding site (21-22). p53 also promotes the expression of some genes through elements that are of limited similarity to the consensus binding motif (e.g., *PIG3*, *PAC1*) (23-25); therefore, sequence pattern discovery algorithms alone cannot reliably predict where p53 will interact with its chromosomal targets nor does the presence of a consensus sequence itself determine whether the site will be occupied *in vivo* by p53. An added complication is that the nuclear concentration of p53 increases one to two orders of magnitude, from a few hundred molecules per cell to perhaps a few thousand of tetramers per cell, in response to certain genotoxic and non-genotoxic stresses. Furthermore, post-translational protein modifications and the presence of other binding partners and their concentration all are thought to modulate p53's ability to transcriptionally activate or conversely repress target genes. Considerable effort will therefore be needed to map the global binding distribution of p53 in mammalian cells.

For our studies we are treating human lung tumor A549 cells with adriamycin for 15 hr and then carrying out standard ChIP enrichment of the cross-linked DNA using D01 as the anti-p53 antibody. After the cross-links were reversed and the repaired DNA ends were ligated with the adapter shown above, limited PCR was used to amplify the fragments with cassette-specific primers, then the DNA was digested with *Bbs* I and cloned into pBEST. Purified plasmid DNA from this clone pool was digested with *Mme* I, and the protruding 3′ ends were removed by incubation with T4 DNA polymerase and deoxynucleotide triphosphates. After blunt-end ligation to form circles, the sample was electrophoresed on a low melt agarose gel and the monomeric circle band was recovered and electroporated into electrocompetent D1210 cells. The cells were plated on ZYM5052 agar plates, and plasmid DNA was prepared as above. These molecules now have the following paired-end diTAG structure:

```
               TAG (18-19bp)                         TAG (18-19bp)
Vector—CGACNNNNNNNNNNNNNNNNNN (N) (N) NNNNNNNNNNNNNNNNNNGTCG—Vector
   DNA——GCTGNNNNNNNNNNNNNNNNNN (N) (N) NNNNNNNNNNNNNNNNNNCAGC-DNA
```

Each tag is 18 (or 19) bp long which, in most cases, is sufficient to allow the site from which the fragments were derived to be uniquely positioned on the genomic map. In practice, since it can be hard to tell just from inspection where one tag ends and the next begins, tags of only 18 nucleotides are extracted.

INCREASING TAG LENGTH AT THE 3′ END

With the strategy described above, the two unique bp at the 3′ end of each tag are lost, which results in an inability to uniquely identify a tag's location in large genomes (Figure 1). One strategy for capturing these nucleotides in the tag is based on the approach used in the process called TALEST (tandem arrayed ligation of expressed sequence tags) developed by Spinella et al. (26) and modified by our laboratory for our original GST protocol (8). It employs ligation with a 16-fold degenerate oligonucleotide to capture all the sequence information in the *Mme* I′ site's 3′ extensions. To further simplify downstream processing of the data, we designed the linker to contain tandem copies of a *Bcl* I recognition site (5′-pTGATCACGTGATCANN-3′). After it is ligated to the *Mme* I 3′ overhangs, digestion with *Bcl* I leaves a single cohesive GATC overhang on the end of each tag, and these linear DNAs can now easily be ligated to form circles. Because *Bcl* I cutting is blocked by *dam* methylation, the enzyme will not cut in the tags or in the vector as the plasmid DNA was prepared from *E. coli* D1210, a *dam*$^{+}$ strain. After ligation and purification of the monomeric circles by agarose gel electrophoresis, the DNA is treated as before by being electroporated into electrocompetent D1210 cells, and the library is plated on ZYM5052 agar plates. The resulting plasmid DNAs now have the following structure:

```
        TAG (20-21 bp)          TAG (20-21 bp)
Vector-CGACNNNNNNNNNNNNNNNNNNNN (N) TGATCA (N)
        NNNNNNNNNNNNNNNNNNNNGTCG-Vector
  DNA-GCTGNNNNNNNNNNNNNNNNNNNN (N) ACTAGT (N)
        NNNNNNNNNNNNNNNNNNNNCAGC-DNA
```

Each tag is 20 (or 21) bp long with the *Bcl* I recognition sequence serving as a clear punctuation mark to divide diTAGs into their respective left and right ends. It is also easy to tell if a tag is 20 or 21 bp long. In principle, several additional linkers based on the above *Bcl* I paradigm could be used provided the cognate methylase is available, e.g., *Bam*H I or *Eco*R I.

*BSE*R I DIGESTION TO RELEASE PAIRED-END DITAGS

Approximately 5-10 μg of plasmid DNA is prepared from plate scrapings and then digested using *Bse*R I as the manufacturer's conditions (NEB). Since each diTAG is flanked by a suitably positioned *Bse*R I recognition site, digestion releases one diTAG pair from every DNA circle. These are the only *Bse*R I sites in the vector, and the paired-end diTAGs can be easily purified from the linearized vector on a 1.5% low melt agarose gel and then concatemerized as described previously for Long-SAGE tags. After size fractionation, the concatemers are cloned back into pBEST cut with *Bse*R I and dephosphorylated to form the paired-end diTAG library. We routinely plate out this library on non-inducing agar plates, e.g., 2xYT, and then pick colonies into 96-well cultures using ZYM-5052 liquid autoinduction medium. Dilutions (1 to 10) of the overnight cultures are boiled for 10 min. to release DNA, which is then used as template in PCR reactions to amplify the concatemer inserts. After incubation with alkaline phosphatase and exonuclease I, the samples are sequenced using the same primers as were used for the PCR reactions. The concatemers have the following architecture if the degenerate *Bcl* I linker was used:

GTCGAC-Tag1-TGATCA-Tag1′-GTCGAC-Tag2-TGATCA-Tag2′-
GTCGAC-Tag3-TGATCA-Tag3′-GTCGAC-Tag4-TGATCA-Tag4′. . . .etc.

Each diTAG pair begins and ends with the sequence GTCGAC (a *Sal* I site), and in-between each set of paired-end tags is a single copy of the *Bcl* I recognition sequence (TGATCA), which makes parsing of the 20 or 21 bp tags straightforward.

PAIRED-END PROFILING OF THE METHYLOME

Because the degenerate linker strategy maximizes the information content at the 3′ ends of the tags, it has become the core strategy for our ongoing analysis of p53 binding sites, and it also is being adapted for global analysis of alterations in the human genome involving 5′ methylation of cytosine in CpG dinucleotides. These alterations are regarded as epigenetic as they control gene expression in cells and during development but do not change the DNA sequence. Seventy percent of all cytosines in CpG dinucleotides in the human genome are methylated and prone to deamination, resulting in a cytosine to thymine transition, CpG to TpG or CpA on the complementary DNA strand (27-28). This process is believed to have led to an overall reduction in the frequency of guanine and cytosine in the human genome to about 40% of all nucleotides and a further reduction in the frequency of CpG dinucleotides to about a quarter of their expected frequency (29). The exception to CpG under representation in the genome is within CpG islands, which were originally called HTFs, for *Hpa* II tiny fragments that remained uncut after digestion with the 5 mC sensitive restriction enzyme *Hpa* II (CCGG) (29). CpG islands were later formally defined as sequences >200 bp in length with a GC content >0.5, and a CpGobs/CpGexp (observed to expected ratio based on GC content) >0.6 (29-30). However, more recent studies have shown that CpG islands located near transcription start sites are usually longer than 500 bp while those less than 500 bp tend to be associated with repetitive elements (31-32).

Determining the global pattern of DNA methylation, or the methylome (33), and its variation in cells is an area of considerable interest because of its potential use as an early diagnostic biomarker for cancer (34-35). Tumor cells exhibit hypomethylation of their genomes, but the promoters of certain tumor suppressor genes (e.g., p16^{ARF}) frequently are silenced in tumor cells through hypermethylation (reviewed in 36). Accordingly, numerous approaches are being developed to identify methylation-silenced or demethylation activated genes. In one approach we are taking, total genomic DNA is digested to completion with *Mse* I (T/TAA), whose recognition site is found rarely within CpG islands but occurs about once every 140 bp in bulk DNA. DNA fragments with methylated cytosines in the digest are then separated from the remainder of the genomic fragments by affinity chromatography (37-39). The methyl CpG fragments can be ligated with

Cassette #2

```
                                         BtgZ I        Bbs I
5′        TCCGGTCTAC  TGAATTCCGA  ACGCGATGCT  GAAGACCACG  AC
3′  Amino-AGGCCAGATG  ACTTAAGGCT  TGCGCTACGA  CTTCTGGTGC  TGATp
```

and then digested with *BtgZ* I and *Bbs* I as in the ChIP protocol. After cloning and *Mme* I digestion, the resulting paired-end diTAGs have the following structure

TAG (20-21bp) **TAG** (20-21bp)
Vector-CGAC**TAA**NNNNNNNNNNNNNNNNN (N) TGATCA (N)
NNNNNNNNNNNNNNNNN**TTA**GTCG-Vector
DNA-GCTG**ATT**NNNNNNNNNNNNNNNNN (N) ACTAGT (N)
NNNNNNNNNNNNNNNNN**AAT**CAGC-DNA

with the nucleotides in bold coming from the *Mse* I recognition sequence. In this case, as shown in Figure 1, tag length is critical since the first 3 bases are already fixed by the remainder of the *Mse* I recognition sequence. Decreasing tag length by trimming off the 3′ extensions after *Mme* I cutting would inflict a sizable penalty on the chances of the tags being unique. Another example of tag size counting is shown in Figure 3, which illustrates the basic principles of the Methylated CpG island Amplification (MCA) protocol developed by Issa and co-workers (40) and how it can be modified to provide paired-end diTAGs. In this case the DNA is first digested with *Sma* I, which only cleaves leaving blunt ends provided the central CpG dinucleotide in its recognition sequence (CCC/GGG) is unmethylated. These methylated sites, however, can be cleaved with *Xma* I (C/CCGG G) to leave a 4 base overhang. Ligation of the overhang with the DNA adapter cassette #3 shown below followed by cleavage with *BtgZ* I or *Bbs* I places 5′ CGAC 3′ overhangs on the ends of what were methylated CCCGGG sequences in the genome.

Cassette #3

```
                                          BtgZ I       Bbs I
5' TCCGGTCTAC          TGAATTCCGA  ACGCGATGCT  GAAGACCACG  A
3' Amino-AGGCCAGATG  ACTTAAGGCT  TGCGCTACGA  CTTCTGGTGC  TGGCCp
```

About 70-80% of CpG islands contain at least two closely spaced (≤1 kb) *Sma* I sites. If they are consecutively methylated they can be used for cloning the intervening CpG-rich segments since after *BtgZ* I and/or *Bbs* I digestion they will have the CGAC overhangs needed for ligation into the *Bbs* I-digested pBEST vector. During cloning the two *Mme* I recognition sequences flanking the inserts are recreated, and the 3° C in the overhang now becomes the last residue in the *Mme* I recognition site. Therefore, cutting with *Mme* I will generate tags that are CGGG plus 16 or 17 nt, which maximizes their information content for determining where these fragments map in the genome (see Figure 1).

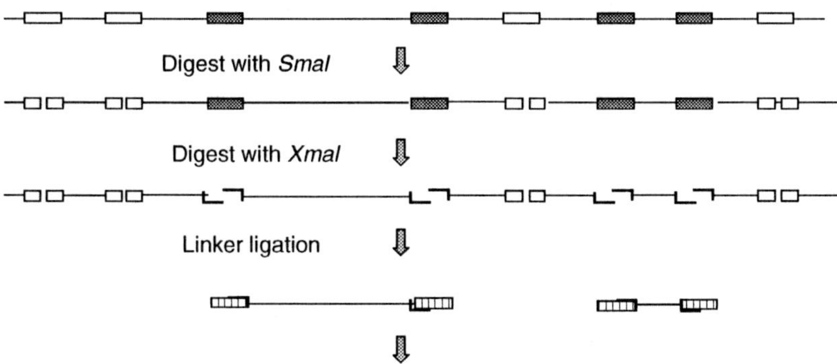

Figure 3. Schematic diagram of *Smn* I/*Xma* I double-digestion protocol. Genomic DNA is represented by a solid line with eight *Sma* I CCCGGG recognition sequences; four sites are non-methylated (open boxes), and four are methylated (filled boxes). The non-methylated sites are cut in a first digestion with methyl-sensitive *Sma* I leaving blunt ends. A second digestion is performed using the methyl insensitive *Sma* I isoschizomer *Xma* I, which leaves CCGG overhangs. DNA adapters with appropriately positioned *BtgZ* I and *Bbs* I sites are ligated to the overhangs, and DNA fragments with an adapter at each end are PCR amplified using primers complementary to the adapters.

SUMMARY

Because paired-end genomic signature tags are sequenced-based, they have the potential to become an alternate tool to tiled microarray hybridization as a method for genome–wide localization of transcription factors and other sequence-specific DNA binding proteins. As outlined here the method also can be used for global analysis of DNA methylation. One advantage of this approach is the ability to easily switch between different genome types without having to fabricate a new microarray for each and every DNA type. However, the method does have some disadvantages. Among the most rate-limiting steps of our PE-GST protocol are the need to concatemerize the diTAGs, size fractionate them and then clone them prior to sequencing. This is usually followed by additional steps to amplify and size select for long (≥500) concatemer inserts prior to sequencing. These time-consuming steps are important for standard DNA sequencing as they increase efficiency ~20-30-fold since each amplified concatemer can now provide information on multiple tags; the limitation on data acquisition is read length during sequencing. However, the development of new sequencing methods such as Life Sciences' 454 new nanotechnology-based sequencing instrument (41) could increase tag sequencing efficiency by several orders of magnitude (≥100,000 diTAG reads/run), which is sufficient to provide in-depth global analysis of all ChIP PE-GSTs in a single run. This is because the lengths of our paired-end diTAGs (~60 bp) fall well within the region of high accuracy for read lengths on this instrument. In principle, sequence analysis of

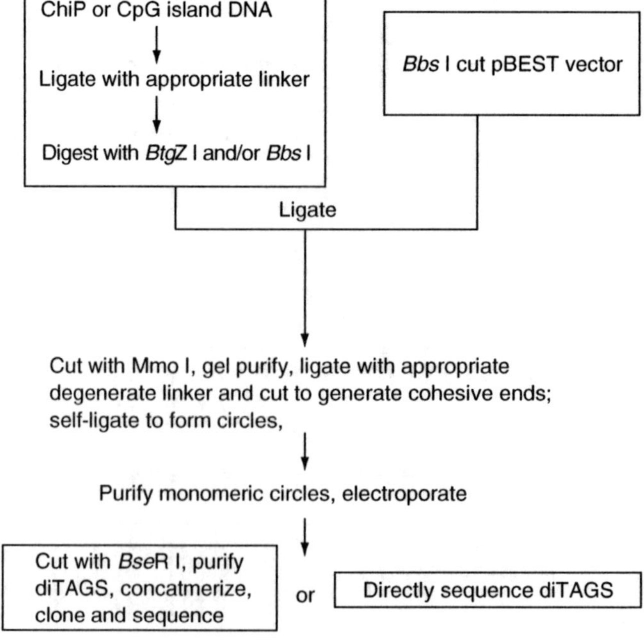

Figure 4. Schematic diagram of the diTAG protocol.

diTAGs could begin as soon as they are generated, thereby completely bypassing the need for the concatemerization, sizing, downstream cloning steps and sequencing template purification. In addition, our protocol places any one of several unique four-base long nucleotide sequences, such as GATC, between each and every diTAG pair, which could be used to help the instrument's software keep base register and also provide a well-located peak height indicator in the middle of every sequence run. This additional feature could permit multiplexing of the data by simultaneous sequencing of several pooled libraries if each used a different linker sequence during diTAG formation (Figure 4).

ACKNOWLEDGMENTS

Support from National Institutes of Health Grant AI056480, the Low Dose Radiation Research Program of the Office of Biological and Environmental Research (BER), U.S. Department of Energy and the BNL Laboratory Directed Research and Development Program is gratefully acknowledged.

REFERENCES

1 Orlando, V. (2000) Trends Biochem. Sci. 25, 99-104.

2 Ren, B., Robert, F., Wyrick, J.J., Aparicio, O., Jennings, E.G., Simon, I., Zeitlinger, J., Schreiber, J., Hannett, N., Kanin, E., Volkert, T.L., Wilson, C.J., Bell, S.P. and Young, R.A. (2000) Science 290, 2306-2309.

3 Wells, J., Graveel, C.R., Bartley, S.M., Madore, S.J. and Farnham, P.J. (2002) Proc. Nat. Acad. Sci. U.S.A. 99, 3890-3895.

4 Saha, S., Sparks, A.B., Rago, C., Akmaev, V., Wang, C.J., Vogelstein, B., Kinzler, K.W. and Velculescu, V.E. (2002) Nat. Biotechnol. 20, 508-512.

5 Impey, S., McCorkle, S.R., Cha-Molstad, H., Dwyer, J.M., Yochum, G.S., Boss, J.M., McWeeney, S., Dunn, J.J., Mandel, G. and Goodman, R.H. (2004) Cell 119, 1041-1054.

6 Tuzun, E., Sharp, A.J., Bailey, J.A., Kaul, R., Morrison, V.A., Pertz, L.M., Haugen, E., Hayden, H., Albertson, D., Pinkel, D., Olson, M.V. and Eichler, E.E. (2005) Nat. Genet. 37, 727-732.

7 Bird, A.P. (1986) Nature 321, 209-213.

8 Dunn, J.J., McCorkle, S.R., Praissman, L.A., Hind, G., Van Der Lelie, D., Bahou, W.F., Gnatenko, D.V. and Krause, M.K. (2002) Genome Res. 12, 1756-1765.

9 Velculescu, V.E., Zhang, L., Vogelstein, B. and Kinzler, K.W. (1995) Science 270, 484-487.

10 Mayr, B. and Montminy, M. (2001) Nat. Rev. Mol. Cell Biol. 2, 599-609.

11 Chen, J. and Sadowski, I. (2005) Proc. Nat. Acad. Sci. U.S.A. 102, 4813-4818.

12 Kim, J., Bhinge, A.A., Morgan, X.C. and Iyer, V.R. (2005) Nat. Methods 2, 47-53.

13 Labhart, P., Karmakar, S., Salicru, E.M., Egan, B.S., Alexiadis, V., O'Malley, B.W. and Smith, C.L. (2005) Proc. Nat. Acad. Sci. U.S.A. 102, 1339-1344.

14 Roh, T.Y., Cuddapah, S. and Zhao, K. (2005) Genes Dev. 19, 542-552.

15 Ng, P., Wei, C.L., Sung, W.K., Chiu, K.P., Lipovich, L., Ang, C.C., Gupta, S., Shahab, A., Ridwan, A., Wong, C.H., Liu, E.T. and Ruan, Y. (2005) Nat. Methods 2, 105-111.

16 Studier, F.W. (2005) Protein Expr. Purif. 41, 207-234.

17 Wahl, G.M. and Carr, A.M. (2001) Nat. Cell Biol. 3, E277-286.

18 Vousden, K.H. and Lu, X. (2002) Nat. Rev. Cancer 2, 594-604.

19 Yang, A., Kaghad, M., Caput, D. and McKeon, F. (2002) Trends Genet. 18, 90-95.

20 Oren, M. (2003) Cell Death Differ. 10, 431-442.

21 el-Deiry, W.S., Kern, S.E., Pietenpol, J.A., Kinzler, K.W. and Vogelstein, B. (1992) Nat. Genet. 1, 45-49.

22 Hoh, J., Jin, S., Parrado, T., Edington, J., Levine, A.J. and Ott, J. (2002) Proc. Nat. Acad. Sci. U.S.A. 99, 8467-8472.

23 Contente, A., Dittmer, A., Koch, M.C., Roth, J. and Dobbelstein, M. (2002) Nat. Genet. 30, 315-320.

24 Kim, E. and Deppert, W. (2003) Biochem. Cell Biol. 81, 141-150.

25 Yin, Y., Liu, Y.X., Jin, Y.J., Hall, E.J. and Barrett, J.C. (2003) Nature 422, 527-531.

26 Spinella, D.G., Bernardino, A.K., Redding, A.C., Koutz, P., Wei, Y., Pratt, E.K., Myers, K.K., Chappell, G., Gerken, S. and McConnell, S.J. (1999) Nucl. Acids Res. 27, e22.

27 Coulondre, C., Miller, J.H., Farabaugh, P.J. and Gilbert, W. (1978) Nature 274, 775-780.

28 Bird, A.P. and Taggart, M.H. (1980) Nucl. Acids Res. 8, 1485-1497.

29 Bird, A. (2002) Genes Dev. 16, 6-21.

30 Gardiner-Garden, M. and Frommer, M. (1987) J. Mol. Biol. 196, 261-282.

31 Ponger, L., Duret, L. and Mouchiroud, D. (2001) Genome Res. 11, 1854-1860.

32 Ponger, L. and Mouchiroud, D. (2002) Bioinformatics 18, 631-633.

33 Feinberg, A.P. (2001) Nat. Genet. 27, 9-10.

34 Baylin, S.B., Herman, J.G., Graff, J.R., Vertino, P.M. and Issa, J.P. (1998) Adv. Cancer Res. 72, 141-196.

35 Marsit, C.J., Kim, D.H., Liu, M., Hinds, P.W., Wiencke, J.K., Nelson, H.H. and Kelsey, K.T. (2005) Int. J. Cancer 114, 219-223.

36 Feinberg, A.P., Ohlsson, R. and Henikoff, S. (2006) Nat. Rev. Genet. 7, 21-33.

37 Cross, S.H., Charlton, J.A., Nan, X. and Bird, A.P. (1994) Nat. Genet. 6, 236-244.

38 Shiraishi, M., Chuu, Y.H. and Sekiya, T. (1999) Proc. Nat. Acad. Sci. U.S.A. 96, 2913-2918.

39 Rauch, T. and Pfeifer, G.P. (2005) Lab. Invest. 85, 1172-1180.

40 Toyota, M. and Issa, J.P. (2002) Methods Mol. Biol. 200, 101-110.

41 Margulies, M., Egholm, M., Altman, W.E., Attiya, S., Bader, J.S., Bemben, L.A., Berka, J., Braverman, M.S., Chen, Y.J., Chen, Z., Dewell, S.B., Du, L., Fierro, J.M., Gomes, X.V., Godwin, B.C., He, W., Helgesen, S., Ho, C.H., Irzyk, G.P., Jando, S.C., Alenquer, M.L., Jarvie, T.P.,

Jirage, K.B., Kim, J.B., Knight, J.R., Lanza, J.R., Leamon, J.H., Lefkowitz, S.M., Lei, M., Li, J., Lohman, K.L., Lu, H., Makhijani, V.B., McDade, K.E., McKenna, M.P., Myers, E.W., Nickerson, E., Nobile, J.R., Plant, R., Puc, B.P., Ronan, M.T., Roth, G.T., Sarkis, G.J., Simons, J.F., Simpson, J.W., Srinivasan, M., Tartaro, K.R., Tomasz, A., Vogt, K.A., Volkmer, G.A., Wang, S.H., Wang, Y., Weiner, M.P., Yu, P., Begley, R.F. and Rothberg, J.M. (2005) Nature 437, 376-380.

INDEX

T

Printed in the United Kingdom
by Lightning Source UK Ltd.
115828UKS00007B/11

9 780387 338408